T0075034

THE IRRESISTIBLE ATTRACTION OF GRAVITY

The mystery of gravity has captivated us for centuries. But what is gravity and how does it work? This engaging book delves into the bizarre and often counter-intuitive world of gravitational physics. Join distinguished astrophysicist Professor Luciano Rezzolla on this virtual journey into Einstein's world of gravity, with each milestone presenting ever more fascinating aspects of gravitation. Through gentle exposure to concepts such as spacetime curvature and general relativity, you will discover some of the most curious consequences of gravitational physics, such as black holes, neutron stars and gravitational waves. The author presents and explains one of the most impressive scientific achievements of recent times: the first image of a supermassive black hole. Written by one of the key scientists involved in producing these results, you'll get a behind-the-scenes view of how the image was captured and discover what happens to matter and light near a black hole.

Luciano Rezzolla is the Chair of Theoretical Astrophysics and Director at the Institute for Theoretical Physics in Frankfurt, Germany. His main research topics are the physics and astrophysics of compact objects, such as black holes and neutron stars. He is a member of the Event Horizon Telescope Collaboration (EHTC), where he sits on the Executive Board. He has received numerous prizes including the Karl Schwarzschild Prize, the Frankfurt Physics Prize, the Golden Seal of the University of Bari, the 2020 Breakthrough Prize for Fundamental Physics (with EHTC) and the Einstein Medal (with EHTC). Since 2019 he has been the Andrews Professor in Astronomy at Trinity College, Dublin.

"What are 'black holes' and do they exist in our Universe? In his well-written and easy to understand account Prof. Luciano Rezzolla explains to the non-expert reader the basic theoretical ideas and the evolution of the scientific research over the past century. He then reveals how in the last few years we have been able to actually identify these weird but fascinating objects through very high resolution imaging with radio waves, as well as the detection of gravitational waves. This is a good read from a top expert in the field."

Prof. Reinhard Genzel, Max Planck Institute for Extraterrestrial Physics; Nobel laureate in Physics

"In recent years there's been a real surge in our knowledge of black holes and their role in the cosmos. Luciano Rezzolla clearly explains the new results, their contexts and the future prospects for research. Having himself been involved in the intricate computer modelling and imaging, he conveys his enthusiasm to the reader through his personal perspective on what it's like to participate in these important discoveries."

Prof. Martin Rees, University of Cambridge; author of *Gravity's Fatal Attraction*

"Luciano Rezzolla offers an engaging overview of the powerful role of gravity, as the weakest but most consequential interaction shaping our universe. The narrative is engaging and scientifically accurate, with up-to-date details at the forefront of astrophysics and fundamental physics. Overall, Rezzolla offers the unique gift of a comprehensive, yet pedagogical summary of the latest exciting developments, such as imaging black holes and the use of gravitational waves as a new messenger across the cosmos."

Prof. Avi Loeb, Harvard University; author of *Extraterrestrial: The First Sign of Intelligent Life Beyond Earth*

"Black holes are mysterious objects. Some of their secrets have now been revealed, not least due to the work of this author. In his book, he describes this fascinating story in an understandable, even entertaining, yet scientifically exact way. You will not stop reading until you have reached the final page!"

Prof. Dr. Claus Kiefer, University of Cologne; author of *Quantum Gravity and Gravitation*

"Why do things fall? Starting from this simple question, Rezzolla takes us on a whirlwind tour of gravity, from Galileo and Newton at the birth of modern science, to Einstein's 1915 revolution of warped spacetime. We are brought right up-to-date with the latest news on gravitational waves and black hole imaging, covering a lot of ground in an engaging style. Rezzolla makes the most complex topics accessible to both non-experts and those wanting to become experts."

Prof. Geraint Lewis, Sydney Institute for Astronomy; author of *The Cosmic Revolutionary's Handbook*

The Irresistible Attraction of Gravity

A Journey to Discover Black Holes

LUCIANO REZZOLLA
Goethe-Universität Frankfurt Am Main

Shaftesbury Road, Cambridge CB2 8EA, United Kingdom

One Liberty Plaza, 20th Floor, New York, NY 10006, USA

477 Williamstown Road, Port Melbourne, VIC 3207, Australia

314–321, 3rd Floor, Plot 3, Splendor Forum, Jasola District Centre, New Delhi – 110025, India

103 Penang Road, #05-06/07, Visioncrest Commercial, Singapore 238467

Cambridge University Press is part of Cambridge University Press & Assessment, a department of the University of Cambridge.

We share the University's mission to contribute to society through the pursuit of education, learning and research at the highest international levels of excellence.

www.cambridge.org
Information on this title: www.cambridge.org/9781009198752

DOI: 10.1017/9781009198776

Front cover image provided by Ziri Younsi

English edition first published 2023

Printed in the United Kingdom by TJ Books Limited, Padstow Cornwall

A catalogue record for this publication is available from the British Library.

ISBN 978-1-009-19875-2 Hardback

To Emilia and Domenico
The undisputed origin of everything

As long as there is imagination, there will be questions
As long as there are questions, there will be hope

CONTENTS

THE BEGINNING OF THE JOURNEY

Gravity attracts – this is such an obvious phenomenon that writing this book was not necessary to stress it. Less obvious is that, even before it appears in the form of physical interaction, gravity attracts our attention and our imagination. As soon as we are born, before developing a conscious relationship with the physical universe, we already know gravity at an instinctive level. For the rest of our lives, it will represent the only one of the four fundamental interactions of which we will have conscious awareness. And from which we will often try to escape.

I wrote this book with the desire to explain what gravity is and why – even though at a subconscious level – we are irresistibly attracted to it. To do this, I plan to take you on a journey through its observable effects and, in particular, into those regions of physics that have been revealed to us by Einstein's revolutionary theory of gravitation, namely general relativity. It will be a virtual journey of course, which will lead us to a place without borders – the realm of fundamental questions about which humanity wonders. But, like all travels, this one also has ambitions of enriching our vision of the world, extending our horizons and, finally, making us realise that we have learned something. For me, writing this book certainly represented all of this.

On the route that I propose, I will do my best to steer us away from the treacherous waters of erudition and technicalities. Instead, we will head out to open seas, making intuition our reference and imagination our guiding star. Along the way, we will make some stops; partly to catch our breath (given the number of new concepts I will present in each chapter), but above all to find the calm necessary to answer some simple yet not trivial questions, such as:

> *Why does an apple fall from the tree instead of floating in space? What is spacetime? What does its curvature consist of, and how is it produced? Can time be bent? How does a black hole work, and how can we 'build' one? How is it possible to photograph it if it does not emit light? What are gravitational waves, and why are they difficult to measure?*

As with any trip, it is good to be prepared for what awaits us and to pack in our luggage things we might need along the way. So, I will bring with me all I have learned in 30 years dedicated to the study of gravity and, in particular, to those aspects that are inextricably linked to the astrophysics of black holes, neutron stars and gravitational waves. The lessons I have learned in these decades have led me to a variety of predictions and discoveries. The lastest achievements, working together with the Event Horizon Telescope Collaboration, have been the publication of the first images of the supermassive black holes at the centre of the M87 galaxy and of the Milky Way.

As for your luggage, it can be lighter and should contain just two essential elements: an abundance of imagination and a good deal of patience. The first will help you to find the answers that 'we in the trade' can easily read from the equations. On the other hand, the second will be helpful because not everything I write about will be immediately clear (although I can guarantee that it will be correct), and not everything you read will seem obvious, perhaps not even reasonable. However, if you arm yourself with imagination and patience, you can be sure you will find answers to each of the questions posed above, and you will come to understand the roles that spacetime and curvature play in explaining what gravity really is – this mysterious force by which we are all attracted.

Starting our journey from an irrational and neonatal instinct will lead us to the shores of pure amazement. Here, we will come to understand what gravity really is and how some of its most bizarre expressions, such as neutron stars and black holes, actually work.

This will be our journey, from instinct to wonder, in Einstein's revolutionary universe.

1 GRAVITY ... ATTRACTS!

As I have already discussed, the title of this chapter is meant to be less trivial than it might seem.

Here, I do not want to simply stress the existence of a physical 'force' exerted between two objects with a mass, which attracts them to each other even if they are at a great distance. Indeed, in Chapter 3, we will discover that this idea, however widespread and easy to understand, is incorrect and, at least in part, misleading. Rather, with this title, I wish to emphasise that there is something – gravity – that attracts us in a broad sense and, above all, attracts our attention. In addition to physical objects, it applies an irresistible grip on our imagination; it can push our thoughts towards radically new horizons, different from the usual ones, and expose us to scenarios that extend far beyond our everyday experience.

But let's proceed step by step. To better define what gravity is, it helps to break down the knowledge we have into three distinct but interconnected levels. In particular, we could say that we have: *instinctive knowledge*, *rational knowledge* and, finally, an *imaginative knowledge* of gravity.

Let's see what they are and how they differ from each other.

Instinctive Knowledge

We all know that instinct is the natural and intrinsic tendency (that is to say, not requiring the intervention of reasoning or reflection) to carry out a particular behaviour. An example of instinct would be what drives us to flinch to protect our heads when we are surprised by a sudden, loud and unknown sound.

When put in these terms, it is difficult to believe that something instinctive or irrational can possibly bind us to gravity. Yet it is so.

Anyone who has handled a newborn will surely have seen first-hand the so-called Moro reflex, named after the Austrian paediatrician Ernst Moro (1874–1951). It is one of the main neonatal reflexes and is widely used in evaluating the function of the central nervous system. To induce this reflex, it is sufficient to take a newborn, even in the first few seconds after birth, lift it in a horizontal position and (making sure it is totally safe) create the sensation of a free fall! The newborn's response to the unexpected and apparent loss of support is a reaction of surprise, accompanied by the sudden opening of the arms and hands in search of a handhold, as is shown in Figure 1.1.

From a medical point of view, the appearance of this reflex is important evidence of the perfect physiological functioning of the infant's central nervous system. That is why we have all been subjected to this test. Any parent who has attended – this has happened to me three times – is well aware of the relief they felt on seeing their son or daughter respond as they should to this rather bizarre stimulus.

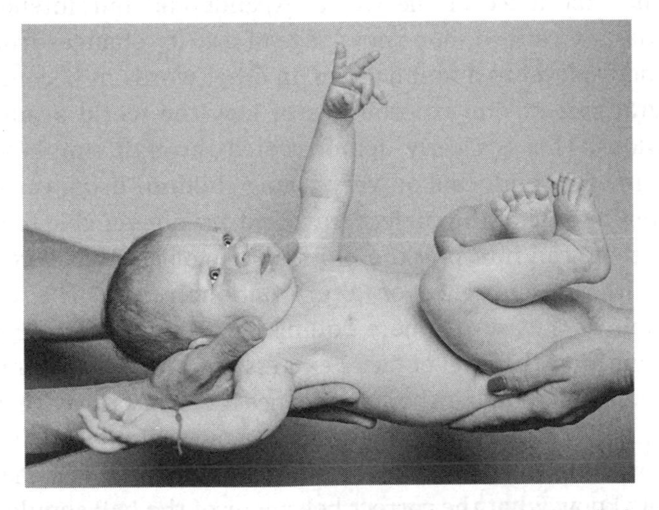

Fig. 1.1 Example of the Moro reflex: a newborn reacts instinctively to the loss of support, spreading her arms and trying to grab a handhold to stop her fall. © V. Tverdokhlib/depositphotos.com.

On an anthropological level, the Moro reflex reminds us of our past as primates, when in all probability we had to be immediately ready to follow our mothers, holding on firmly to their backs. However, what interests us most here is its physical significance. The presence of this instinct a few seconds after birth (when we are completely defenceless and know nothing about the world around us) reveals an important truth concerning our interaction with gravity: we know it instinctively, well before having conscious interactions with the rest of the physical universe. Thus, after having spent nine comfortable months in our mothers' wombs, essentially isolated from everything, we are immediately able to respond to gravity (or rather, to its absence). This is no small detail.

The Moro reflex disappears at around six months, so our knowledge of gravity, while remaining partly instinctive, changes over time as we develop the ability to observe the physical universe and understand its laws.

Rational Knowledge

As our experience of the world expands and our intellectual abilities are refined, our knowledge of gravity changes from an instinctive level to a rational one. In other words, it becomes an integral part of our expectations of how the world around us functions. This is clearly demonstrated through simple visual experiments conducted on very young children using cartoons. Although some of the subjects involved are not yet able to walk, all of them can interpret the motion of an object that is consistent with the presence of a gravitational field. The classic example is that of a sphere rolling on a table: children react differently (at the level of facial expressions and eye movements) depending on whether the sphere, once it reaches the edge of the table, falls, or continues its motion despite being without support or, even more bizarrely, if it begins to fly. The fact that the infants know what the correct behaviour of the ball should be is further confirmation of how deeply the knowledge of the 'force' of gravity is rooted in our minds.

This plays a fundamental role in our rational perception of reality. Thanks to this profound influence, our brains are able to solve very complex dynamic problems in a very short time and with practically no effort. A simple example is the action of hurrying down a flight of stairs: it is one of the most complex problems when programming robots (and one at which they often fail); however, we humans can cope without even consciously thinking about it. Yet, establishing in what sequence and at what speed our movements must be performed to ensure the subtle balance between the various forces competing with the gravitational force, is by no means trivial.

Finally, another property of gravity is worth pondering: its ability to stimulate our imagination.

Imaginative Knowledge

If it is clear that we have an instinctive and rational notion of gravity, then in my opinion it is equally clear that gravity also exerts an irresistible attraction on our imagination. It is precisely because we are immersed and subjected to a gravitational field every instant of our lives that we are naturally fascinated by those scenarios in which gravity is weak or even absent. Who has never wished they could jump off a high cliff or a mountain top and . . . fly? Who has never imagined being an astronaut aboard the International Space Station or a character in a science-fiction movie floating effortlessly from one place to another in zero gravity? I often do . . .. In other words, gravity attracts our attention. It stimulates our imagination precisely because it is the only fundamental 'force' of which we have a conscious awareness and because we know all too well how difficult it is to escape it. What, if not imagination, led first Isaac Newton and then Albert Einstein to explain (in very different ways) the laws that govern gravity?

There are many examples that illustrate the powerful pull of gravity on our imagination, but I will limit myself to just one, which I find representative and easy to understand. In 2013, the Spanish director Alfonso Cuarón made a movie with the emblematic title: *Gravity*. For almost two hours, the film talks about nothing other than gravity; or rather, its absence. However, not many know that *Gravity* broke the box office record for the first weekend screening of the films released that autumn. Some may argue that this success is entirely due to the leading actors, the two Hollywood superstars Sandra Bullock and George Clooney. In my opinion, however, a fundamental role has been played by the fact that – like it or not – we cannot escape gravity and its irresistible pull on our minds.

One of Four, but Definitely Different from the Others

The considerations made so far provide an excellent starting point for introducing another important reflection on gravity's role within our understanding of nature. Modern physics teaches us that there are four types of fundamental interactions, which essentially describe all the processes taking place in the universe: the *electromagnetic interaction*, the *strong interaction*, the *weak interaction* and the *gravitational interaction*.

The first, the electromagnetic interaction, among other things, allows you to read this book regardless of the format you are using. In fact, electromagnetic waves (photons or, more simply, light) emanate from the page in front of you and reach your eyes. These waves then convert into electrical signals transmitted via the optic nerve to the brain, which (thanks to a complex combination of electrical and chemical exchanges) translates them into the words you are reading. The electromagnetic interaction is also responsible for the cohesion and dynamics of the molecules that compose us. Without it, we would not even exist as human beings, and our molecules would scatter like bits of paper in the wind. The theory that describes the electromagnetic interaction is well known, both at the level of classical physics (where it is expressed by the Maxwell equations) and at the level of quantum physics when dealing with elementary particles (where it is also known as theory of *quantum electrodynamics*).

The second, the strong interaction, develops instead on the smallest scale accessible to us in nature, of the order of a few fermi (or femtometres), that is, a few thousandths of a billionth of a millimetre. The strong interaction is a hundred times more intense than the electromagnetic interaction and is exerted between *quarks*, the constituent parts of elementary particles such as protons or neutrons. In reality, this interaction is also present on slightly larger scales, that is, inside atomic nuclei (which generally have dimensions of the order of 10 fermi), where it takes the name of the *strong nuclear force*. In the case of protons and neutrons, the mediating particles of the strong interaction are called *gluons*, while in the strong nuclear force they are represented by the *pions*. In essence, the strong interaction acts as the 'glue' that holds the atomic nuclei together, from the smallest atoms (hydrogen) to the larger ones (for example, uranium). It also regulates what happens when two protons smash against each other at a speed almost equal to that of light or when a neutron star is born (a phenomenon we will discuss in detail in Chapter 5). The theory that describes this interaction is well developed and is called *quantum chromodynamics*. Unfortunately, given the complexity of the theory and the equations that describe it, it is often difficult to make precise predictions, especially if the energies are high or if the number of particles involved is large, as in the case of neutron stars.

The third, the weak interaction, is responsible for the radioactive decay of some atomic nuclei and acts between *leptons* (a class of particles to which electrons, certainly the most 'familiar' of the group, belong) and quarks. It is thanks to this interaction that neutrinos (very light particles produced by matter of high density and temperature, such as those at the centre of the Sun) interact only rarely ('weakly') with protons and neutrons, the ordinary matter of which we are composed. It is good to recall that, at this precise moment, our bodies are crossed by billions of neutrinos emitted about eight minutes ago by the Sun and which travelled to us at almost the speed of light. If we do not 'feel' pierced by these particles, it is precisely because they interact only weakly with the ordinary (or *hadronic*) matter of which we are made. In other words, there is little to worry about; we are essentially transparent to neutrinos. The theory that describes the weak interaction is also very well developed and can be combined with that of electromagnetism, so in that case we refer to it as the *electroweak interaction*.

So, finally, we arrive at the fourth, the gravitational interaction. Although it is not yet the time to explain in detail what it consists of and how it links to one of the most subtle and elegant concepts of theoretical physics – the curvature of spacetime – we can already reflect on what distinguishes it from the others. Gravity is the only physical interaction of which we are consciously aware. The only one whose influence on your body you are aware of at this moment, as you read this book. In fact, whether you are lying on a bed, slumped in an armchair or standing, *you know* that some 'force' affects your position. Without it, you would levitate, floating freely just like an astronaut on the International Space Station.

This point is essential in order to fully understand the title of the chapter and grasp its deeper meaning: gravity attracts our attention because we can directly and tangibly experience its existence, unlike other fundamental interactions. By contrast, we cannot appreciate the level of cohesion of the molecules that make up our bodies, how rarely we interact with a neutrino or how many radioactive particles we emit.[1]

The fact that we can experience it directly and consciously is enough to give gravity a special place among the fundamental interactions and, in my opinion, places it above all others. What makes it even more peculiar is that this awareness accompanies us in every second of our lives, even before coming into the world. Subconsciously, we are aware of the presence of gravity long before grazing our knees when we learn to walk and run, and certainly before we encounter the laws of physics in school or college.

But What Is Gravity, and How Does It Work?

In all likelihood, many of you feel convinced you can provide a reasonable answer to the simple questions: 'What is gravity?' and 'How does it work?'. This is because you know it both instinctively and rationally, and (either at school or university) you have been given an 'educated' explanation for it. However, it is equally likely that the explanations given to you are not entirely correct, even if they are not completely far-fetched either. What I mean is that what you have been told or taught isn't really *wrong*, but it's not *right* either.

The reason for this apparent contradiction is that it is possible to understand gravity on various levels. For example, the representation of gravity proposed by Newton is simpler and more intuitive, which involves the existence of a *gravitational force* and whose mathematical description is relatively simple. At the same time, a different and deeper understanding is possible: the one suggested by Einstein, which implies a geometric vision of space and time and a much more complex mathematical description.

In the following chapters, a sort of evolution awaits us; we will develop our understanding of gravity. The first stage of this evolution – and therefore the first destination of our virtual journey – will take us from the basic understanding of gravity as it is encoded in our brains, even on an instinctive level, to the description provided by Newton's theory. The second stage – or the next destination of the journey – will instead lead us to the mathematically compact and physically profound description Einstein proposed with his theory of general relativity, elegantly embodied in his field equations.

Thus, we will learn to connect what we know and experience on our planet, where gravitational fields are very weak, to what we observe in the universe, where enormously stronger gravitational fields give life to marvellous phenomena such as black holes, neutron stars and gravitational waves. In this way, we will find that our knowledge and understanding of gravity is strongly influenced by the way it manifests itself on this planet.

However, we will also realise that we have to abandon such a view because it is wrong and because the constraints it sets on our imagination are too tight.

All in all, the correct answer to the questions posed above will be clear: 'Gravity is simply the manifestation of the curvature of spacetime'. At the moment, this statement is still cryptic, I know. But, as mentioned in the introduction, a certain amount of patience is needed to embark on the journey that awaits us. I can assure you that it will all become much clearer by the end of Chapter 3.

2 THE FATHERS OF GRAVITY

To fully understand gravity, it may be helpful to take a small step back in time and adopt a historical perspective rather than a physical one. This interlude will be rather short and limited to those scientists who, more than others, have contributed to shaping our understanding of the subject.

Galileo Galilei: The Importance of the Method

Let us start from the beginning. To paint a historical portrait of the theory of gravity, and thus acknowledge those who its 'fathers' were, we cannot help but start in Italy, in particular in Pisa. There, between the end of the sixteenth century and the beginning of the seventeenth century, Galileo Galilei (1564–1642) was the first to try to decode the dynamics (i.e., the state of rest or motion) of physical objects. These were still difficult times for science: deductive logic and experimental pragmatism inevitably succumbed to philosophical and religious dogmatism. Galileo experienced this first-hand when the accusation of heresy was levelled at him in 1633.

Nonetheless, adopting *the scientific method* that he himself had introduced, Galileo intended to study quantities, such as the velocity and the state of motion at constant velocity. In this way, he empirically derived the so-called *principle of inertia*, which, thanks to Isaac Newton, became an axiom of dynamics more than a century later.

This principle states that, in the absence of external forces, a body maintains its state of uniform rectilinear motion (i.e., it continues to move at a constant speed along a straight trajectory) and remains at rest if it is initially at rest. The meaning of this principle is less detached from reality than it may seem; we experience its effects every time we are on a bus or the subway and the vehicle starts or brakes. In the first case, we are accelerated in the opposite direction to that of travel simply because we tend to maintain a state of rest. In the second case, however, we are accelerated in the direction of travel since our 'inertia' compels us to maintain the state of motion we had before. Thus, we continue to move forward.

This discovery led Galileo to formulate the *Galilean principle of relativity*, from which the so-called *Galilean transformations between two reference systems* follows. In essence, it states that the laws of physics are the same in two frames of reference (or, equivalently, for two observers) that are moving at constant or zero speed. As we will see, this principle was also useful to Einstein about three centuries later, and that is why Galileo can be considered the 'first relativist'.

However, besides objects in uniform motion, Galileo was also very interested in objects subject to forces and, therefore, experiencing acceleration. The simplest way to study objects in accelerated motion is to let them fall, for reasons that will soon become clear. Therefore, Galileo spent most of his time on top of towers (like the one in Pisa, and those in Bologna and Florence), concentrating his attention on the motion of 'gravi' (from the Latin *gravis*, or heavy), that is, of heavy objects in free fall.

A practical problem that Galileo faced in his experiments was the fairly high gravitational acceleration of physical objects and, although he exploited relatively tall buildings, the time taken by those objects to reach the ground was still very short and, therefore, difficult to measure. He then resorted to a device that allowed him to slow down the fall time as he pleased: the so-called *inclined plane*. This is a right-angled triangle with the hypotenuse facing upwards; by varying the *attack angle* (i.e., the acute angle), it is possible to vary the component of the 'force' of gravity involved in the motion. By exploiting this simple but ingenious solution, Galileo was able to slow down the motion of heavy spheres rolling down a plane that he had carefully polished to reduce friction. Thus, he was able to perform very accurate measurements.

In this way – and as explained in his most important work, the *Dialogue Concerning the Two Chief World Systems*, published in 1632 – Galileo calculated that the gravitational acceleration (that is, the acceleration of a body subject to the action of gravity) was equal to about 9.80 metres per square second (m/s^2). This estimate is essentially correct, which is quite surprising if one thinks of early seventeenth-century technology and the precision it allowed. Today, we know that this acceleration varies according to the position on the surface of the Earth, with values ranging between 9.764 m/s^2 and 9.834 m/s^2, and changes in relation to altitude. However, the standard reference value of 9.80665 m/s^2 is still very close to Galileo's measurement. Furthermore, Galileo estimated that the distance travelled by a free-falling object increases with the square of the time taken to fall – a result we now know to be correct.

To this day, historians still wonder whether Galileo really did conduct experiments such as dropping objects of different material (wood, gold and silver), or of the same material but with different masses (cannonballs and musket balls), from the top of the Tower of Pisa. One thing is certain, however: in his *Dialogue* he summarised a series of considerations (both his own and those of previous scientists) pointing to the evidence of a universal law of gravitation. This hypothesis, which Galileo never formulated in explicit terms as a law, essentially states that all objects in a gravitational field accelerate in the same way, regardless of their mass or composition. In other words, Galileo had already guessed that if you throw a cannonball and a musket ball from a high tower, they will reach the ground simultaneously.

This important and deep hypothesis – which today we call the *equivalence principle* and know to be correct with a relative accuracy of one-millionth of a billion – is particularly valuable for two distinct reasons. The first is that it openly clashed with Aristotle's assertion, namely, that gravitational acceleration is directly proportional to the weight of an object, so that heavy ones should fall faster than light ones. It is difficult for us, with our modern scientific mindset, to imagine the intellectual effort and courage needed to take sides against the cultural authority represented by the Aristotelian dogmatic principle at the beginning of the seventeenth century. The second is that to believe (contrary to our direct experience) a cannonball and a feather in free fall accelerate in the same way required a considerable leap of imagination. Indeed, Galileo's conclusion is far more revealing than it may appear and points to a consideration of fundamental importance: the evidence gained through our daily experience can be misleading! More specifically, the reason for which, in our everyday experience, the feather reaches the ground after the cannonball is not because the feather experiences a smaller gravitational acceleration but because it is subject to proportionally larger friction than the cannonball. If dropped from a tower in a vacuum (therefore, in the absence of drag), the feather and the cannonball would touch the ground simultaneously (this was indeed demonstrated at the end of the last Apollo 15 moon walk and can be watched on public archival videos). Obviously, Galileo could not create vacuum conditions in his experiments from the Tower of Pisa, but his imagination was powerful enough to show him the underlying truth. In the following pages, we will see that a similar feat of imagination will be made by Einstein almost three hundred years later.

Galileo's results and, above all, the first use of a scientific method (*his* very own method!) to abandon a metaphysical interpretation of reality in favour of a deductive experimental approach were undoubtedly essential for what was discovered and understood about gravity after him. However, at the time of his death in 1642, knowledge on the subject did not stray far from pure empirical evidence. It took about 40 years for a new and more complete view of gravitational interaction to emerge.

Isaac Newton: A Complete Mathematical Picture

The first mathematical formulation of the theory of gravitation is indisputably associated with Isaac Newton (1642–1726), who consequently may rightly be counted among the fathers of gravity. There is little doubt that the contributions of this brilliant mind to the construction of modern science have been enormous and wide-ranging: from astronomy to mathematics; from theology to alchemy ..., to illustrate them, if only briefly, I would need to write another book. But that's not my aim here. What I want to focus on, instead, is Newton's contributions to the understanding of gravity contained in what is considered his main work: *The Mathematical Principles of Natural Philosophy* or, more simply, the *Principia* (from the original Latin title: *Philosophiae Naturalis Principia Mathematica*).

In this three-volume treatise, published in 1687, Newton introduced three principles of the dynamics of bodies that represent the backbone of the so-called Newtonian physics. Although well known and almost emblematic of our scholastic knowledge of physics, it is helpful to recall them briefly:

1. *Principle of inertia*: in the absence of forces, a body will maintain its state of rest or uniform rectilinear motion.
2. *Principle of conservation of momentum* (given by the product of mass and velocity): in the presence of a force, the change in motion (i.e., the acceleration) is proportional to the resulting force applied and occurs along the direction in which the force is exerted.
3. *Principle of action and reaction*: for every action of one body upon another, there is a corresponding equal and opposite reaction.

Alongside these three principles (and taking gravity as belonging to a particular class of forces), Newton also proposed the so-called *universal law of gravitation*. The basic properties of the force of gravity postulated by Newton can then be summarised as follows:

1. *It is an attractive force between two massive bodies*. In other words, this force manifests itself as an attraction between two bodies with mass (i.e., a property that can be associated with any material physical body) along the direction that joins them.
2. *It is an instant force*. In other words, the two bodies experience this force immediately so that if something changes in one of

the two (the mass or the position, for example), the other body 'knows' about it instantly.

3. *It is a force proportional to the masses of the two bodies involved.* Hence, the greater the mass of the bodies, the greater the intensity of the force. Note that this force does not depend on the composition of the bodies themselves or on their size. The only thing that matters is how massive they are. A direct consequence of this property is that a body without a mass is not subject to a gravitational force.
4. *It is a force inversely proportional to the square of the distance between the two bodies.* Stated differently, once the mass is fixed, bodies at a great distance attract much less than bodies that are close to each other. It follows that bodies at an infinite distance do not attract each other at all; similarly, the same body would experience a gradually weaker force as it moves away from a centre of gravitational attraction.

When writing this book, I made a deliberate choice to limit mathematics to the bare minimum but not to do without. When used at the right moment, mathematics can help us understand the laws of physics, even when used only descriptively. In this spirit, we can use mathematical expressions – the same ones Newton used over three centuries ago – to summarise what is expressed in the four points above. In essence, those four statements can be condensed into this 'conceptual equation':

$$\begin{pmatrix} \textit{strength of the} \\ \textit{force of gravity} \end{pmatrix} = \frac{(\textit{mass of object } 1) \times (\textit{mass of object } 2)}{(\textit{distance between object } 1 \textit{ and } 2)^2}$$

The most important aspect of the expression above (which should be readable even for those who are 'allergic' to mathematics) is the 'equals' sign. Indeed, the $=$ symbol, which we are all familiar with, makes the above expression an *equation*, thus establishing that what appears on its left-hand side is equivalent to what is written on its right-hand side. However, if we want to translate all this into an actual mathematical equation, then the universal law of gravitation is expressed by the equation:

$$\vec{F}_{grav} = -G\frac{M_1 M_2}{r^2}\left(\frac{\vec{r}}{r}\right) \tag{2.1}$$

Here, M_1 and M_2 are the masses of the two bodies, while r is the distance between them. The small arrows above $\vec{F}_{grav}$ and $\vec{r}$ remind us that they express quantities with a strength and a direction. These are also called *vectors*: the vector of the gravitational force and the vector of the distance between the two bodies, respectively.[1] Finally, the symbol G simply represents a constant of proportionality between what is written on the right and what appears on the left; it has no deep physical meaning and must be determined through experiments. Today, we know that, according to the International System of Units, $G = 6.67408 \times 10^{-11}\mathrm{m}^3/(\mathrm{kg\,s}^2)$.

The striking and fascinating aspect of the simple Equation (2.1) is that, with it, Newton was able to provide a satisfactory and extremely accurate description of gravitation. Thanks to this mathematical tool, it was possible to interpret any phenomenon that had something to do with gravity: to explain why an apple falls from a tree or why the Earth orbits the Sun. Combined with the ideas put forward by the Polish astronomer Nicolaus Copernicus (1473–1543) (namely, that all the planets close to us revolve around the Sun) and the German astronomer Johannes Kepler (1571–1630) (that orbital motion occurs along elliptical trajectories with the Sun positioned in one of the two foci), Newton's universal law of gravitation (Equation (2.1)) could explain all the astronomical observations of his time. By virtue of this, and for about two centuries, it was possible to make extremely accurate predictions about the motion of the celestial objects in our solar system.

Another remarkable property of Newton's expression is that it is 'universal'; in other words, it is all we need to explain, at least approximately, *any* gravitational phenomenon on Earth. Whether we want to build the longest bridge or the tallest skyscraper, we need nothing more than this simple and beautiful Equation (2.1). Nonetheless, such an equation and the concepts on which it is built are *incorrect*. More specifically, we are now aware that:

1. *Gravity is not a force*. The concept of 'gravitational force' isn't only not useful but also misleading. Therefore, it should be replaced by another concept, which is better suited to explain the nature of gravity; however, such a concept is more complex to express mathematically, and therefore farther from our intuitive understanding of nature.
2. *Gravity is not instantaneous*. Modern physics is built on the opposite assumption: every physical interaction propagates with a finite speed, the fastest of which is the speed of light. Gravity too is subject to this restriction, so a change in the gravitational field has to propagate over a finite time before it is detected.
3. *The mass alone is not enough to describe gravity*. In addition to the magnitude of the mass, its distribution in space must be considered. In other words, it is necessary to know the length-scale over which the mass is contained. In the case of an extended object such as the Sun, the properties of gravity relate as much to its mass as to its radius.

For over two centuries, Newton's law of gravitation was used extensively, and its undisputed validity was confirmed by an endless series of experimental successes, thus making it, among other things, the pillar of navigation and, therefore, of maritime trade. However, a simple astronomical observation was sufficient to undermine a theory that until then seemed written in the stone of universal laws and was treated with an almost reverential attitude.

A small crack opened up in the noble and imposing facade of the building Newton had erected, conjuring up its inevitable collapse: *the precession of the perihelion* of Mercury. This seemingly enigmatic name refers to a very easily observable sequence of events. Every 88 days or so, Mercury makes a complete orbit around the Sun, reaching the point in the orbit when it is closest to the star, that is, the *perihelion*. In Newton's theory, this point is always the same, fixed and invariable unless external factors intervene. In essence, Mercury's orbit around the Sun must be

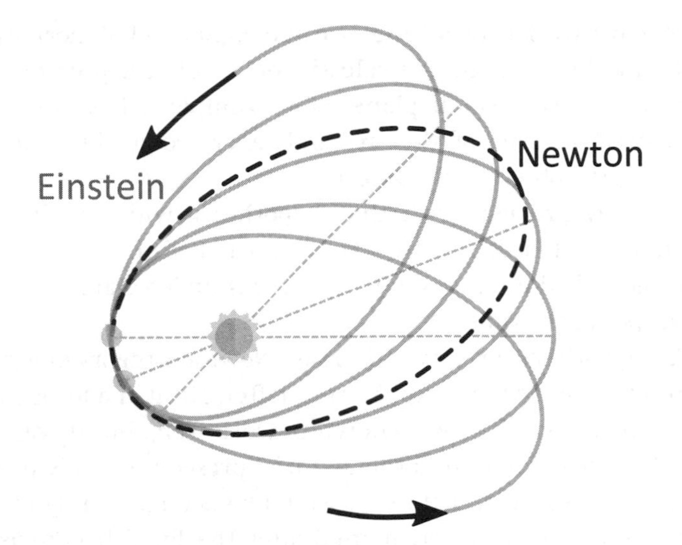

Fig. 2.1 Schematic illustration of Mercury's orbit. According to Newton's theory of gravitation, it should be closed (dashed ellipse), while Einstein's theory of gravity (solid line) tells us that it must continuously change, with the perihelion moving at each orbital period.

'closed', that is, it must return to exactly the same point after an orbital period. This is shown in Figure 2.1, where the dashed ellipse indicates Mercury's orbit according to Newton's law of gravity.

However, astronomical observations tell us something entirely different! Even after considering the orbital variations due to other bodies in the solar system, Mercury's perihelion changes more than Newton predicted, that is, by 0.1035 seconds of an arc (or arcseconds) each orbital period, which corresponds to 0.4297 arcseconds per year.[2]

For decades, there were attempts to provide an explanation for the bizarre anomaly shown by the motion of Mercury that would be compatible with Newton's theory of gravitation. And what is perhaps even more surprising (and will give us the opportunity for an important comment at the end of the chapter) is that the astronomers of the time managed to reconcile the irreconcilable, that is, to explain the anomalous motion of Mercury through Newton's law of gravitation! However, the price to pay (which, over time, became higher and higher) was to postulate the presence of a celestial object of very precise mass and orbit, a very small planet, for example, whose presence influenced the orbit of Mercury and generated, at the appropriate time, the observed deviation.

The only problem with that hypothesis, which in itself is not too far-fetched, was that the 'planet' had to be effectively invisible: it had to be so small that it was undetectable with the telescopes of the time.

The hypothesis put forward to save Newton's theory of gravity represented a *plausible* explanation (after all, it made use of a theory of gravity that was known to work well), but it was also an *unlikely* one: it required the artificial presence of a disturbing object never observed before. It would be enough to rely on the principle of *Occam's razor*, named after the English Franciscan philosopher and friar William of Ockham (1288–1347), to consider this solution as unreasonable. Hence, when several hypotheses are equally plausible to explain a certain physical behaviour, the iron logic of Occam's razor requires that the simplest be considered the correct one. Obviously, nature does not always choose the simplest route, so Occam's razor serves, above all, to discard the most cumbersome hypotheses. However, Occam's razor was never employed with Mercury's orbit. On the contrary, in the absence of a better theory of gravity and given the deference to Newton's theory, the 'little planet' hypothesis could satisfy almost everyone.

Albert Einstein: A Revolutionary Vision

More than two centuries after the publication of Newton's *Principia*, the world was offered a new theory of gravity. Not only was it even more general, providing a natural explanation to the motion of Mercury, but also its predictions on the precession of perihelion coincided perfectly with the observations. This theory, which initially appeared incomprehensible even to the most illustrious scientists of the time, destroyed the profound certainties about gravity acquired with Newton and proposed a radically different vision. So different that even today it is not without some difficulty that we come to accept its implications. I am talking about general relativity: a prodigious result of the mind of Albert Einstein (1879–1955).

General relativity rightly represents one of the most important scientific theories ever formulated and is a supporting pillar of modern physics. Without it, most of the observations from particle physics, astrophysics and cosmology would be inexplicable.

Published at the end of 1915, over the last hundred years general relativity has passed a long series of experimental tests, and today it is the theory of gravitation that best matches astronomical observations and laboratory experiments. However, it represents an even more impressive success if we consider that general relativity remains one of the most mathematically complex theories ever formulated in physics. The solution of its equations still represents a challenge.

But it is not yet time to go into the details of this theory: we will do this in the next chapter, where we will come to understand how important and far-reaching its implications are.

In the meantime, let's return to the mysterious phantom planet that was necessary to explain the orbit of Mercury within Newton's theory of gravitation. We have seen that such an explanation was possible and even plausible, but not likely. It is precisely this fact that prompted more daring minds to find less-cumbersome solutions. From this point of view, the evolution of the theory of gravity (from Galileo to Newton and then to Einstein) illustrates well a fundamental point: it is essential to view a scientific theory as a logical and mathematical framework built to explain the laws of nature under certain conditions. It provides, within a historical context, a satisfactory explanation of the laws of nature. Therefore, each theory is intrinsically improvable as our knowledge of nature progresses (particularly through observations or experiments that challenge our initial assumptions) or as our mathematical ability to address the weaker aspects of the theory improves.

Hence, the historical developments of gravity described in this chapter teach us that a physical theory, unlike a mathematical theory (based on postulates and logic), is only a temporary representation of a complex manifestation of nature. It is born to be continually improved, revised or completely abandoned if it cannot explain some new and unexpected manifestation of the laws of nature. This fate awaits even the best theory of gravity we know, general relativity, on the facade of which (about a hundred years after its construction) very small cracks have begun to appear.

3 SPACETIME, CURVATURE AND GRAVITY

This chapter has the daring ambition of explaining a series of abstract concepts underlying Einstein's theory of general relativity in a simple way. The mathematics of this theory can express such concepts in a transparent and elegant manner, and, for a physicist or a mathematician, (almost) everything is clear as soon as the corresponding equations are written down. So, I could simply show them and say, 'It's all there, don't you see?'. I know, however, that such an approach would not take us very far, while my commitment is to be your vigilant guide on this journey.

So, what I am going to do is definitely more difficult: to illustrate this new interpretation of gravity using concepts we are all used to. The challenge in this approach is not so much related to the complexity of the concepts (which is all in the mathematics) but with the effort necessary to accept an interpretation of physical reality that is often in contrast to our everyday experience. In other words, what I am asking is that you imagine a reality that can be very far from what you are used to. But on the other hand, imagination is one of the requirements of this journey.

So let us start from where the previous chapter ended, namely, when we said that the Einstein equations elegantly summarise the equivalence between gravity and the curvature of spacetime. To be more precise, we wrote that these equations essentially state that:

$$\left(\begin{array}{c}\textit{gravity is the manifestation of}\\ \textit{spacetime curvature}\end{array}\right) \tag{3.1}$$

I know from experience – and from reading the faces of far too many students when first written on a blackboard – that this sentence sounds fundamentally obscure. However, it is correct and absolutely true, at least in so far as we can verify it through scientific experiments. Do not despair, though, and, above all, do not give up now! Later on, when we return to this statement, you will find yourself able to understand it and even appreciate its profound implications.

Spacetime: Everything That Has Happened and Will Happen

To understand what Equation (3.1) actually states, we can analyse the various terms that it contains. So, let us start from the concept of *spacetime*. First of all, it should be noted that I have not written 'space-and-time', nor 'space–time', but 'spacetime'. It may seem a minimal, irrelevant difference, but it is instead fundamental because it represents one of the first lessons of the theory of general relativity. This is because Einstein postulates that our idea of space and time as distinct entities is incorrect since the two are completely equivalent. It follows that a shift in space from one point to another is equivalent, at the level of the equations, to a shift in time from one moment to another. Going left or right, up or down, is not in any way different, at least mathematically, from going back and forth in time.

In Einstein's theory, therefore, spacetime is a single object, which can be seen in essence as a 'container' of elements (i.e., termed a 'set' in mathematics) that are also referred to as *events*. Let me explain by offering an example: The European Union can be considered a 'set' of citizens from its various countries; such citizens are therefore the building blocks, or elements, of the

European Union. Similarly, spacetime is the set of all events: past, present and future. An event (an element of spacetime) is understood to be something that has happened, is happening or will happen somewhere at some time. The alarm clock you set this morning, having breakfast, taking a bus to work . . . all of these are events. Reading this book right now is also an event, as it is possible to establish that it happens in a specific place (for example, on your sofa) and at a specific time (now). Since each event needs the specification of four different pieces of information – three for the position in space and one for that in time – it is quite simple to conclude that spacetime is a container in four dimensions (three spatial and one temporal) of elements called events.

To this definition, Einstein adds that this container of events is not particularly useful when taken alone and without further rules. Therefore, it is necessary to establish precise rules to identify the relative position of the various elements, that is, their *distance*. In other words, when considering two events A and B, it is essential to establish in what relation they are to each other, namely, whether A occurred before or after B, whether it is to its right or its left, above or below and, finally, whether the two are somehow connected or, conversely, completely independent.

These rules can only be established by making a choice of coordinates, that is, after introducing an ordered series of numbers that establish the sequence of events and thus allow us to know the relative position of A and B. This choice of coordinates is totally arbitrary – hence the adjective *relativity* in Einstein's theory – but the relations among the events that follow from the selection of coordinates is the same, regardless of the choice made. This is why, within the theory of general relativity, it is still possible to establish without ambiguity what you did first this morning: wake up or have breakfast.[1] Furthermore, the sequence of events will be the same for every observer, whatever their state of motion.

Curvature

Now that we understand what spacetime is, let us go back to our statement (3.1) and focus on the second concept that appears there, that of *curvature*.

Again, mathematics offers us sophisticated and elegant tools for measuring the curvature of a surface in a space of arbitrary dimensions. Without going into detail, what we need to know here is that there are surfaces that can be considered 'curved' and that their degree of curvature can be measured using rather simple mathematical tools, one of which is the concept of *parallel transport in a closed circuit*. The way it works is as follows: take a vector (i.e., an 'arrow' of a certain length) in a certain portion of spacetime and move it along a closed loop, making sure it always points in the same direction. If you do this, you are performing a *parallel transport*, and there is a mathematical object (called the *Riemann* or *curvature tensor*) that will measure the amount of curvature present when going around the loop.

This may sound abstract, but a couple of examples can help clarify this concept. Therefore, consider a surface in two dimensions, starting from the simplest possible, that is, a plane. Next, imagine representing the closed circuit by means of a triangle ABC and using, as a reference for a certain direction, an arrow of a given length whose origin coincides with point A and whose direction is aligned along the side AB. In mathematics and physics, an arrow of this type is called a 'vector' and is characterised by a length and a direction indicated by the tip of the arrow. We can now slide our vector along AB, bring it to point B, and move it towards point C. In doing so, however, it is important to keep the initial direction of the arrow fixed since this is the idea behind parallel transport. Once at C, we will transport the vector to the AC side and close the circuit, as shown in Figure 3.1.

At this point, we can compare the final vector with the initial one, concluding that they are identical: nothing has changed in carrying it in parallel along a closed loop contained in a plane. Furthermore, we will say that this absence of a change in the two vectors is linked to the absence of curvature on the surface, that is, that the plane is a *flat* surface because it has zero curvature.

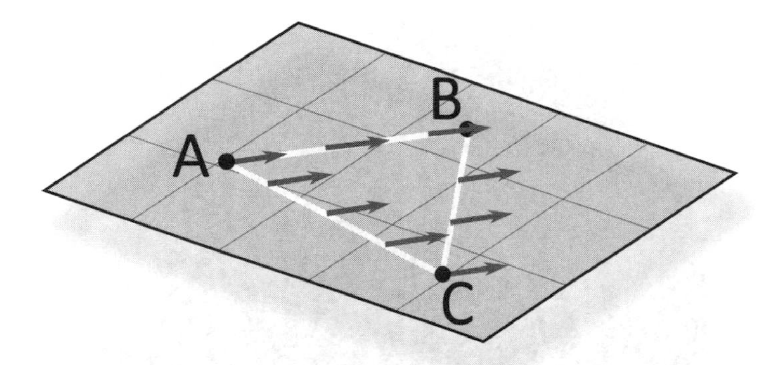

Fig. 3.1 Parallel transport of an arrow (vector) along the triangle ABC on a two-dimensional plane. The comparison between the final vector and the initial one at A shows us that they are identical: the plane is a flat surface because it has zero curvature.

What we need next is an example of a non-flat surface. Let us consider a two-dimensional surface that mathematicians adore for its simplicity and for the richness of its properties: a *two-sphere*, that is, the surface of a sphere with a constant radius. We can think of a two-sphere as the peel of an orange if it were smooth and very thin, or as a simplified representation of the surface of the Earth.

In this case, we could build our closed circuit (again represented by a triangle ABC) by connecting two meridians (lines of longitude) through the section of a parallel (line of latitude) that intersects them. So let us imagine starting from point A, perhaps a beautiful island in the Caribbean, and pointing north with the help of a compass. As long as we move precisely following the needle, we are essentially carrying our vector parallel along the meridian and can therefore reach the North Pole, that is, point B. There, being careful not to turn around, and thus always looking in the same direction, we go south following another meridian, for example, the one placed at a right angle to the first, and we thus reach point C somewhere in northern Africa. Once we arrive at our destination, at the latitude of A, we move west along the parallel that forms the side CA of the triangle but always looking in the direction in which we reach the North Pole.

Once we are back at A and have closed the circuit, as shown in Figure 3.2, we can compare the final vector with the initial one,

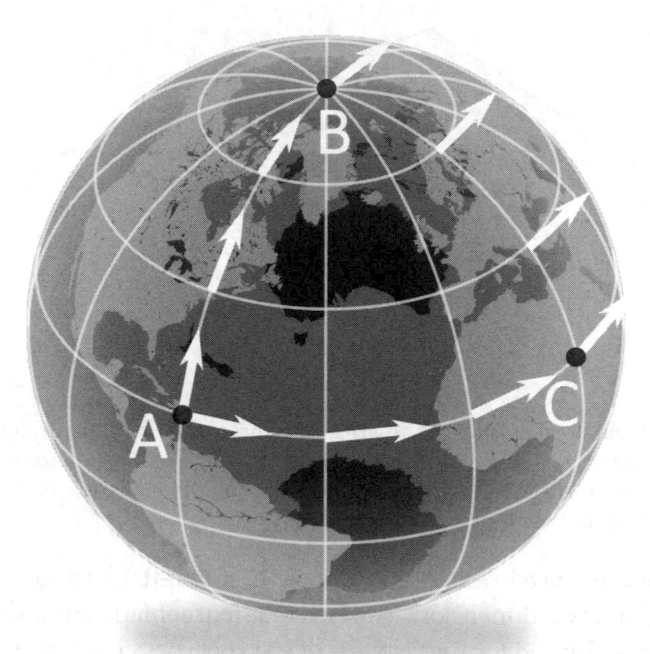

Fig. 3.2 Parallel transport of a vector along a closed circuit on a two-sphere. The comparison between the final vector and the initial one shows that the two are different because they point in perpendicular directions: the two-sphere is a curved surface as it has a non-zero curvature.

concluding that this time they are not identical at all: they point in perpendicular directions, the first to the north and the second to the east!

In this way, we have found that the parallel transport of a vector along a closed loop on a two-sphere introduces a change in the direction of a vector and that this change is related to the presence of a curvature on the spherical surface. In other words, we have learned that a two-sphere is a *curved* surface because it has a non-zero curvature.

Of course, this simple experiment should be enough to convince flat-earthers of how foolish their theory is. But it also reveals something very interesting and invites us to a series of important reflections. The first is that if we had chosen a very small triangle, the variation between the two vectors would have been less evident. For example, if we considered a triangle with sides of just a hundred metres, it would have been very difficult to measure the difference between the vectors without very sophisticated tools. Therefore, a curved surface can be approximated, at least *locally* (i.e., in the vicinity of each of its points), with a flat surface. In other words, the effects of curvature are evident only on scales comparable to the so-called *curvature radius* of the curved surface.

To better understand this point and greatly simplify it, we can say that it is always possible to approximate, at least locally, a curved surface with a spherical one. In this case, the two-sphere that approximates the curved surface – also called *osculating* surface – will have a radius equal to the local radius of curvature of the curved surface. Thus, if we limit ourselves to a surface in a single spatial dimension (that is to say, to a one-dimensional curve), such a curve can always be approximated locally with a circumference whose radius represents the local radius of curvature.

This is illustrated in Figure 3.3, where the black line shows a one-dimensional cut of a generic curved two-dimensional surface. Clearly, when moving from left to right, the curvature along the line varies from point to point; we can surmise though

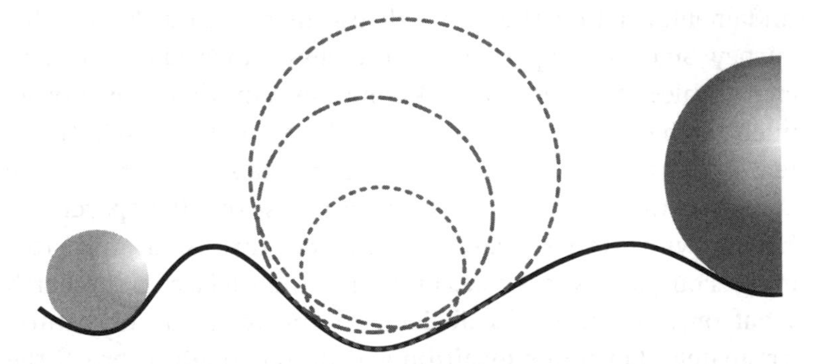

Fig. 3.3 The black line represents a one-dimensional cut of a generic curved two-dimensional surface. The curvature along the line varies from point to point but can always be approximated locally with a circle (a two-sphere in the two-dimensional surface). The radii of the circles can vary and be very large if the line is almost straight.

that, at each point (or region around a point), we can find a circle (a two-sphere in the two-dimensional surface) that approximates, at least locally, the line. It should be equally clear that the radii of these circles are all different and that where the line is locally straight, the circle's radius can become very large and is not shown in the figure.

If, on the other hand, we wanted to consider spacetime in its four dimensions, we would need to employ a four-dimensional sphere (which I am the first to admit is difficult to imagine) osculating at every point the local curvature of four-dimensional spacetime.

Given these considerations, it will come as no surprise that a flat surface (which we have seen is one with zero curvature) still has a radius of curvature associated with it, but this radius is infinite. In the case of the Earth, the curvature radius is equal to its radius, that is, about 6,400 kilometres, which is why it is necessary to consider scales as large as the distance from the Caribbean to the North Pole to measure the difference between the vectors. If we had conducted the experiment in a square as large as the San Marco square in Venice, measuring a difference between the vectors would have been much more difficult, although not impossible with present technology.

The second consideration, triggered by the parallel transport on a two-sphere, follows directly from the first and teaches us a fundamental notion that we will use many times during this journey. Since our experience of the physical world develops on length-scales of a few tens of kilometres (our view does not go much further than that) and is built based on the gravitational field present on Earth, it is almost inevitable to develop the perception that the laws of physics are those of a flat spacetime, that is, with zero curvature. However, this is an incorrect impression, just as it is to assume that the surface of the Earth is flat only because it is difficult to appreciate its curvature. Fortunately, Einstein's intuition was deep enough to break the constraints of our perception of flat spacetime. We will return to this point in the following chapter.

The third consideration is that the mathematical tool employed so far (the parallel transport of a vector along a

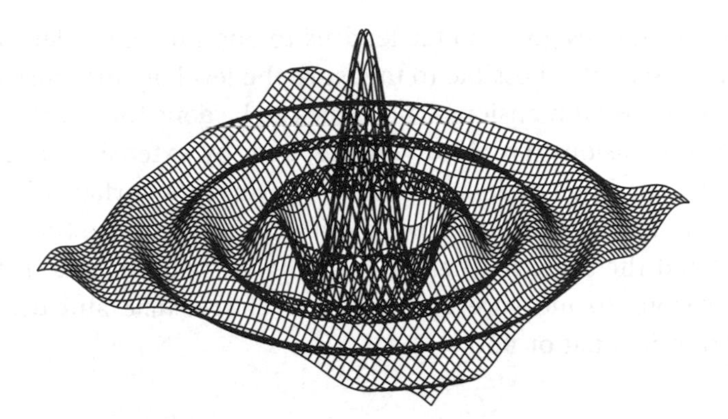

Fig. 3.4 A generic curved surface can be complex and have a local radius of curvature that depends on the position, with a positive or negative value depending on whether the surface is convex or concave.

closed circuit) is entirely general and can therefore also be applied to surfaces whose curvature is much more complex than those considered so far. When looking at Figure 3.4, we can see a two-dimensional surface whose curvature is different from point to point, except in the regions far from the centre. In particular, if you think about the local radius of curvature, it is very small near the origin and becomes larger as you move away from the centre. Exactly at the origin, the radius of curvature is infinite because the function I have chosen in representing the two-dimensional surface (i.e., $\sin(r)/r$) has a constant value of about one near the origin. Finally, the example in Figure 3.4 shows that the curvature can even change sign, becoming positive or negative depending on whether the surface is *convex*, as in the case of the two-sphere, or *concave*.

The theory of general relativity uses sophisticated mathematical tools, many of them developed by mathematicians such as Tullio Levi-Civita and Gregorio Ricci-Curbastro in the early twentieth century, to measure the radius of curvature and its sign. Thanks to them, it would not, for instance, be difficult to develop apps in our smartphones, which would allow us to measure the local curvature of spacetime similar to those apps that measure the local value of your altitude.

The examples given so far lead us to one final consideration: we have seen it is possible to measure the level of curvature of a surface in two dimensions. The same can be done for a surface in three dimensions (i.e., in a volume) and, by extension, in four dimensions (such as in spacetime) or even in a surface of arbitrary dimensions. In other words, it is possible to use the same logic and the same mathematical tools introduced here in two dimensions to measure the *curvature of spacetime* and deduce whether it is flat or not.

Curvature and the Einstein Equations

Now that we have seen what curvature is and which mathematical tools can be used to measure it, we can take an important step forward and discover how the theory of general relativity tightly links the presence of curvature to that of matter or energy. This link is expressed by the Einstein equations proposed in 1915 and marked one of the most important chapters of modern physics. However, before discussing these equations, it is necessary to recall another important equation (beware the singular!), which is also related to Einstein but has nothing to do with gravity. Indeed, it was introduced 10 years before the theory of general relativity was proposed and is part of the so-called theory of 'special relativity' (1905). This well-known equation reads:

$$E = mc^2 \tag{3.2}$$

This expression, also known as the *Einstein equation*, has become an icon of the German physicist's scientific contribution. Although obtained without having gravity in mind, it expresses an important idea that we will frequently use in the rest of the book. In essence, through the equals (=) sign, Equation (3.2) states the equivalence between the energy (E, to the left of the equals sign) and the mass (m, to the right). The presence of the square of the speed of light, c^2, is essentially irrelevant to what we are discussing here and serves only to make the two sides of the equation comparable in terms of units (for this reason, it is shown in grey).[2] In other words, Equation (3.2) tells us that whenever we have a mass (it may be represented by a feather or a boulder, an elementary particle or a star), we also have available a certain amount of energy.

This profound truth is behind the functioning of stars, which produce energy by fusing (through the process of *thermonuclear fusion*) four hydrogen atoms to form one of helium, whose mass is slightly smaller than the sum of the masses of the four initial atoms of hydrogen. This small difference in mass is released in the form of energy. The same principle applies to nuclear power plants where, through the process of *nuclear fission*, a complex atom such as uranium is 'broken down' to create many smaller ones, whose combined mass is less than the original. Again, the difference in mass is released in the form of energy.

This is why, from now on, we will always talk about *mass–energy*, without specifying whether it is a mass, by which I mean as a set of elementary particles (such as electrons, protons or neutrons) to which we can associate a mass at rest, or pure energy.

Now, without further preamble, we can move on to writing down the *Einstein equations* (note the plural in this case!):

$$G_{\mu\upsilon} = \left(\frac{8\pi G}{c^4}\right) T_{\mu\upsilon} \tag{3.3}$$

It is easy to understand how, for the vast majority of you, these equations are as clear as a Chinese text, that is, if you cannot read Chinese! Finding these equations incomprehensible is perfectly normal – you would need to attend a full university course in general relativity to understand them. For example, for the course I give in Frankfurt am Main in Germany, I need the best part of six months of lectures and the introduction of a series of mathematical tools to be able to explain and manipulate these equations in some detail.[3] However, in this case, we can avoid being 'distracted' by the mathematics and rather concentrate on the deeper conceptual meaning of these equations, which represent the cornerstone of the theory of general relativity and are, therefore, at the centre of Einstein's revolutionary vision of gravity.

Let us start once again from the sign of equality in equations (3.3). It expresses an equivalence between what appears on its left-hand side – the *Einstein tensor* $(G_{\mu\nu})$ – and its right-hand side. The latter is made of two pieces: one in brackets, which has no physical meaning and for this reason is shown in grey, and another tensor, namely the *energy–momentum* tensor $(T_{\mu\nu})$. In essence, the Einstein equations (3.3) can also be read as the following conceptual equation:

$$(\mathit{Einstein\ tensor}) = (\mathit{energy\text{–}momentum\ tensor}) \tag{3.4}$$

Since the Einstein tensor represents a measure of the curvature of spacetime (in fact, it is proportional to the Riemann tensor) while the energy–momentum tensor provides a measure of the quantity of mass–energy present in spacetime (the greater the matter and its concentration, the larger this tensor is), the complex equations (3.3) essentially state that:

$$(\mathit{geometry\text{–}curvature}) = (\mathit{mass\text{–}energy}) \tag{3.5}$$

This is arguably the most important concept in the whole book. It expresses exactly what is contained in the Einstein equations (3.3) in a more complex but also mathematically elegant and physically revolutionary manner. In other words: *wherever there is matter (or energy), one can find a spacetime curvature*. This is always true, no matter whether we are talking about the mass of a galaxy or a grain of sand, a light ray or a cannonball. Only the magnitude of this curvature will vary, that is, how pronounced it is. In the next chapter, we will learn how to distinguish curvatures of different strengths and understand how nature is able to produce them. First, however, we need to see how gravity enters into the conceptual framework we have just discussed.

Spacetime Is an Elastic Fabric

Although rich in important implications that we will explore in the rest of the book, Equation (3.5) still does not explain how gravity fits with what has been said so far. In other words, it is probably not clear yet how the Einstein equations, summarised by Equation (3.5), can explain what causes an apple to fall from a tree. This is the next conceptual step we need to take and will require a little more concentration.

To understand how curvature and gravity are inextricably linked, let us start by imagining a flat but elastic and deformable surface, for example, the sheet of a newly made bed. In the absence of weights, it appears perfectly smooth and flat, and so it remains until something disturbs its surface. This is shown in Figure 3.5, which also helps us to understand the meaning of the Einstein equations: the absence of mass–energy leads to a spacetime with zero curvature, therefore flat. Within such spacetime, we can take a vector, transport it in parallel along a closed circuit and, at the end of the process, find that it is identical to the initial one.

To be picky, a spacetime of this type (strictly empty) cannot exist since the presence of matter or energy, even if infinitesimal, would create a perturbation, hence removing the absolute 'flatness'. However, the example is useful nonetheless because it is

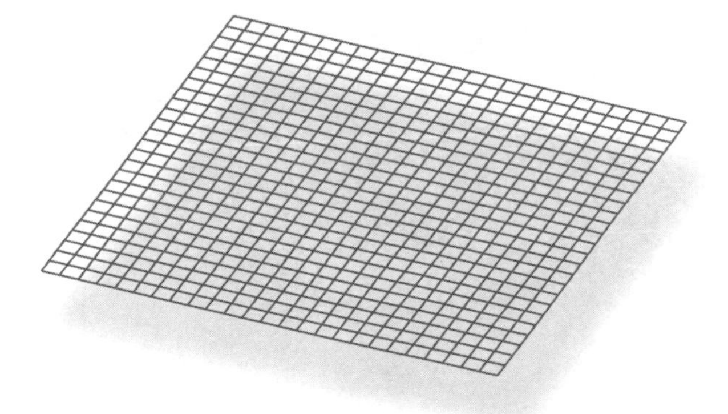

Fig. 3.5 The absence of mass–energy produces a spacetime without curvature, which is, therefore, flat.

often possible to consider a 'nearly flat' or 'locally flat' spacetime that, though not flat on large scales, is flat in a small region.

Let us now try to imagine what would happen if we placed a heavy object on that sheet, such as a bowling ball. As you can easily guess, it would 'curve' around the weight, exactly as shown in Figure 3.6. This example offers us a useful analogy with what is expressed by the Einstein equations: the presence of matter (or energy) produces a spacetime with non-zero curvature and is, therefore, curved. If we took a vector and transported it in parallel along a closed circuit inside it, at the end of the path we would find it is different from the initial one.[4]

The example of the sheet gives us the opportunity for three observations. The first has to do with an essential aspect of Figure 3.6 that should not be missed: the curvature produced by the bowling ball is not identical everywhere but varies from point to point. In particular, it is very small at the edges of the sheet, which will appear almost flat, and more pronounced as it approaches the point where the object is placed, where it will be at its maximum size. In other words, it is natural to expect that, under generic conditions, curvature is a function of position; only in exceptional circumstances (as in the case of a two-sphere) is the curvature constant.

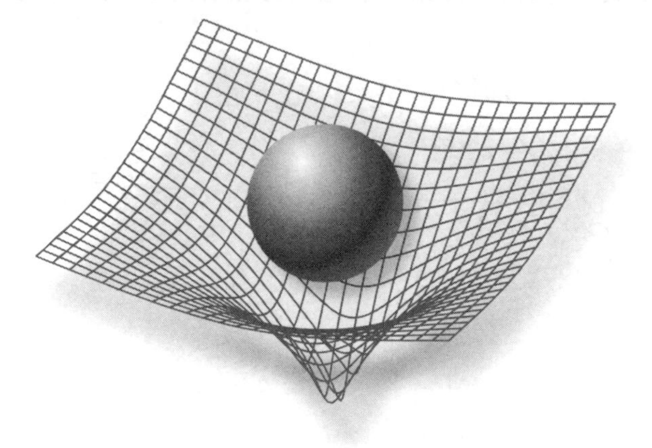

Fig. 3.6 The presence of mass–energy (represented here by a sphere, which could be a planet or a grain of sand) produces a spacetime with non-zero curvature, which is, therefore, curved.

The second consideration concerns the 'shape' of the deformation. Although it may seem obvious that the presence of the bowling ball produces a *concave* or *positive* curvature, the Einstein equations reveal that it is also possible to have, in a completely equivalent manner, a *convex* or *negative* curvature. Our experience is that the matter around us, and with which we normally conduct experiments, actually produces a positive curvature, precisely as illustrated in Figure 3.6. However – and indeed, the Einstein equations allow for it – the existence of matter that produces a *negative* curvature cannot be excluded.

The third and final observation has to do with the impact of the curvature. It is important to appreciate that the presence of a curvature will affect the dynamics (i.e., the motion) of any object around it. For example, let us imagine adding another smaller and lighter sphere next to the bowling ball, such that it does not produce a curvature of the sheet comparable to the first. In such a scenario, and in the absence of friction, the second object will be forced to move, 'falling' towards the centre of the curvature produced by the bowling ball. With this example, we can therefore understand how gravity manifests itself in geometric terms. By modifying the spacetime in which motion occurs, curvature produces a variation in the dynamics of the bodies, which would otherwise remain in the state of rectilinear motion or rest.

Curvature and Gravity

Having introduced and discussed the concepts of spacetime and curvature, we are finally ready to return to the simple but obscure statement (3.1). I recall that it read:

$$\left(\begin{array}{c}\textit{gravity is the manifestation of}\\ \textit{spacetime curvature}\end{array}\right)$$

At this point in my course of general relativity, the bewildered and sceptical looks that appear on the students' faces when they first encounter this statement are replaced by subtle smiles of satisfaction. The expression of someone on realising that: 'Of course! How could it be otherwise?'. Not only is the sentence no longer cryptic, but it has also become deeply fascinating.

In other words, Einstein has stated that if spacetime is empty, in the sense that it is devoid of any matter producing a curvature, then it is also flat. In this case, every object in it will maintain its state of motion: depending on its initial condition, it will continue to be stationary or move with a constant speed along a straight line (this is indeed the principle of inertia). However, if spacetime contains mass–energy producing a curvature, then every object in it will be affected by this curvature through a modification of its state of motion. Thus, those objects initially at rest will no longer be able to remain so. On the other hand, those objects moving at a constant speed along a straight line will see their trajectory and speed change.

It follows that, in this different and decidedly revolutionary view of gravity, if the apple falls from a tree, it is not because there is a gravitational force that draws it towards the Earth. What happens is that the Earth, with its content of matter and energy, curves spacetime around itself and forces objects to move accordingly. When the apple is no longer attached to the tree, which until then prevented it from moving freely, it can only move following the local curvature of spacetime and therefore towards the centre of the Earth. And that is actually what it does! Well, at least until it encounters the planet's surface, which prevents it from continuing further. This is what Figure 3.7 shows in a very schematic (and obviously out of scale!) manner.

An additional comment could be added here, which I am sure you will understand perfectly at this point. The reason, once free from the tree, the apple does not move away into interplanetary space is linked to the *sign* of the curvature, that is, to the fact that the Earth produces a positive (or concave) curvature. However, if we hypothetically admitted the existence of a type of matter producing a negative curvature, then the surface of the 'sheet' in Figure 3.6 would be deformed with a convex curvature, and gravity would behave in the opposite manner: it would be not attractive but *repulsive*, with the apple flying into space! Fortunately, even in the presence of this 'change of sign',

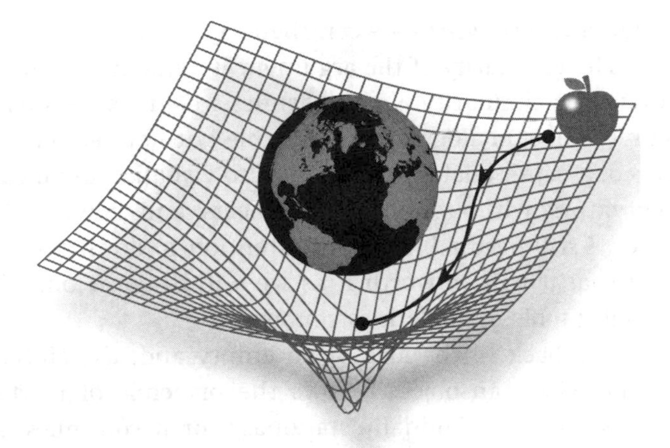

Fig. 3.7 Interpretation of the fall of an apple in Einstein's theory of gravitation: the Earth produces a curvature of spacetime, and the apple must move following it, thus 'falling' towards the centre of the planet.

the title of this book would remain valid. Indeed, in this case, gravity would be even more attractive to our attention!

The example just suggested of a repulsive gravity may seem absurd at first sight, but it is exactly what we think is happening right now to the universe as a whole. By observing very distant supernova explosions, it is possible to deduce that the universe is currently in a phase of accelerated expansion. The cause of this phenomenon is not yet clear, but it could be linked to *dark energy*: a tenuous form of energy that would generate a negative curvature. As a result, at least on a cosmic scale, it would appear to produce an action similar to that of *antigravity*!

Nothing Moves in a Straight Line

We have already discussed in note 4 how the example proposed in Figure 3.6 contains a logical flaw. There, to explain curvature and gravity, we have used a bowling ball already immersed in a gravitational field (the terrestrial one) so that it curves the sheet precisely because it is already in such a field. Therefore, we are using the dynamics of an object in a gravitational field to explain what curvature is, which we can then say is the very origin of gravity. The circularity of the argument is evident. Fortunately, however, this circularity is also harmless: it just shows us that the curved-sheet analogy is very valuable but not perfect.

Indeed, a better analogy can be used to visualise and measure the curvature without the aid of sheets and bowling balls. However, I still need a bit of your attention to do this, but rest assured that it will be worth it as we will discover one of the most useful tools in modern astronomy.

Let us go back to the idea of an empty and, therefore, flat spacetime where an object follows the principle of inertia: it remains stationary if initially stationary, or it continues in its uniform rectilinear motion (i.e., at constant speed). However, it is clear that this can no longer be true if the spacetime is curved. In this case, in fact, the trajectory of the object will be modified by the curvature and will have to adapt accordingly. Hence, uniform rectilinear motion can no longer exist in a curved spacetime, where instead, every motion must necessarily be curved. In other words, nothing moves in a straight line in a spacetime containing mass–energy, except for infinitesimally short distances.

The conclusion just reached has two interesting consequences. The first is related to the common experience that a straight line represents the minimum distance between two points, the so-called *geodesic* between them. This is obviously true in flat spacetime but ceases to be true in a curved spacetime, where a straight line no longer represents the minimum distance.

This consideration is less bizarre than it might appear, and two concrete examples will help us understand how it is quite obvious. First, suppose you are moving along a curved surface and need to go from point A to point B, as shown in Figure 3.8.

You may find yourself in a similar situation when walking on the hills near Trieste in Italy, where these types of deformations of the terrain are very common. Obviously, you would not be able to reach B from A in a straight line (well, unless you fly, but we will consider this case in the next example); instead, you would be forced to follow the surface of the terrain through the route ACB or ADB, choosing the shortest and least tiring

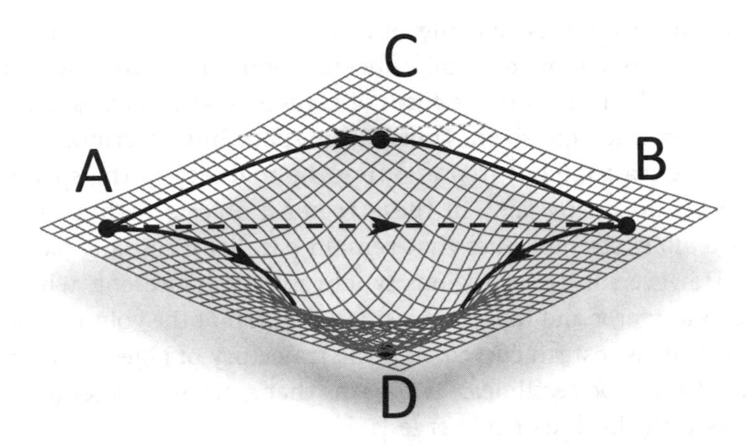

Fig. 3.8 In a curved spacetime, a straight line is not the shortest path to connect two points (i.e., a geodesic). In the figure, the ACB trip is the geodesic between A and B.

one. To clarify, let us suppose it is the first one (although, in different conditions, it could be the second). In this case, the route ACB represents the *geodesic* between A and B, the curve that when travelled measures the shortest distance between the two points A and B.[5]

For the second example, just think of the routes that aeroplanes or ships follow to move from one point to another on our planet. These routes also represent geodesics since they are calculated to minimise fuel costs and, therefore, trace the shortest distance between the two points. However, since planes and ships are still forced to move along the Earth's curved surface, these geodesics cannot be straight lines: they must necessarily be curves, that is, the geodesics of a curved spacetime. This has an important implication: we have seen that it is possible to approximate locally a curved spacetime with a flat spacetime; by the same principle, aeroplanes and ships will seem to move along trajectories that appear as straight lines on small length-scales, but that are actually curved on scales comparable to the radius of Earth's curvature.

The second important consequence of the conclusion that nothing moves in a straight line is indeed linked to astronomy.

Observations tell us that the universe is fundamentally *empty*, with the exception of local concentrations of matter, which manifest themselves in the form of planets, stars, galaxies, clusters of galaxies and so on. Thus, in the spacetime describing our universe, vast regions (representing the majority of the spatial volume) are *nearly flat*. In these regions, objects with a mass move along straight lines that remain parallel if initially parallel. However, in those regions where matter is present, which are much rarer and represent a small fraction of the volume, the spacetime is instead curved, and the trajectory of objects with a mass cannot be rectilinear: two paths that start as parallel must necessarily diverge or converge.

Based on these considerations, we can 'easily' create tools that allow us to verify whether spacetime is, in fact, flat or curved, even without resorting to the aid of sheets and bowling balls. It would be 'enough' to take small objects with mass, such as microsatellites, and launch them towards a fairly distant celestial body, such as Proxima Centauri (which is the closest star to us after the Sun and is only 4.24 light years away), taking care that their trajectories are initially parallel. At this point, it would be sufficient to check the separation between the objects when they have reached their destination, in our case after about 270,000 years (assuming that they are moving at the maximum speed reached so far by an artificial satellite). Depending on whether the separation has decreased or increased compared to the initial one, we will know whether they will have encountered regions of spacetime with positive or negative curvature during the journey.

This example is wildly unrealistic and, although conceptually correct, it is impossible in practice; it is only the abstruse idea of a theoretical physicist. However, a hundred years ago, Einstein suggested an experiment that is quite similar conceptually and was also physically feasible for the technology of those times. Indeed, through that experiment he convinced even the most sceptical scientists of the time about the correctness of his bizarre theory of gravitation.

To better understand what Einstein proposed, we need to go back to his equation on the equivalence between mass and energy, that is, Equation (3.2). That equation allows us to deduce that if, in a curved spacetime, the motion of an object *with a mass* is not rectilinear, the same must apply to the motion of an object *without a mass*, for example, a light ray, which we also refer to as a photon. In other words, a light ray propagating in a curved spacetime will be forced to follow a curved trajectory. At the risk of being pedantic, but with the intent of being extra precise, I should remark that a light ray always moves along a geodesic, a line of minimal separation. This will be a straight line only in a flat spacetime and will instead be a curved line in more general conditions, such as those in a curved spacetime.

When compared to satellites, photons offer two additional advantages: they are much faster than any object with mass (indeed, they travel at the maximum speed allowed); and they are emitted in abundance by celestial bodies such as Proxima Centauri. Therefore, all we have to do is use telescopes to intercept these photons and deduce whether their trajectory has been modified or not by the presence of curvature.

This is exactly what Einstein proposed about a hundred years ago with the famous 1919 total solar eclipse experiment, the basic idea of which is shown in Figure 3.9. In essence, Einstein made a very clever consideration: if a ray of light is deflected by the presence of a massive object (for example, the Sun), which curves spacetime and tends to 'attract' the ray, it is in principle possible to 'see' what lies behind that object!

Obviously, the ability to deflect a light ray depends on the strength of the curvature produced by the celestial body and therefore (as we will see in the next chapter) on its mass and radius. Doing some maths with an approximate version of his equations, Einstein concluded that while it was not possible to observe objects placed *exactly* behind the Sun, the curvature generated by our star would be sufficient to show us objects placed close to its projected surface.

At that point, there was only one problem to solve: the Sun is the dominant source of light around us, and it is therefore impossible to identify the small contribution to the brightness made by a star near its edge. To check this idea, we would need to turn off the Sun! Fortunately for us, this cannot be done, but it

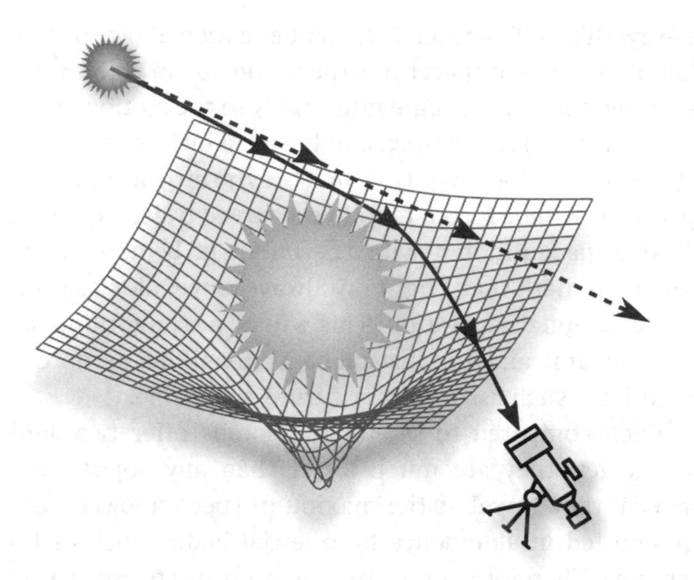

Fig. 3.9 One of Einstein's most important predictions is that light cannot travel in a straight line in a curved spacetime. As a result, it is possible to observe celestial bodies even when they are hidden by other objects in the foreground.

is what essentially happens during a total eclipse. In that case, the light reaching us from the Sun is practically suppressed, so that celestial objects close to its projected surface could be seen – well, at least according to general relativity! The experiment suggested by Einstein, therefore, consisted of observing a total solar eclipse from an optimal position and determining which of the stars close to its edge, and theoretically invisible because covered by the Sun, were actually visible.

I am omitting here the historical details of the adventurous expedition led by the English astronomer Sir Arthur Eddington, a great admirer of Einstein and one of the few initial supporters of the theory of general relativity. Observations made by Eddington on May 29, 1919, on Príncipe Island, at that time a Portuguese territory but now part of the Democratic Republic of São Tomé e Príncipe (an Atlantic archipelago located in the Gulf of Guinea), revealed that it was indeed possible to measure a deflection of light produced by distant stars and that this was equal to about 1.75 arcseconds. The measurement of a physical phenomenon impossible in Newtonian gravity, but that coincided with what had been predicted by the revolutionary theory of general relativity, represented the final straw for Newton's theory and the scientific consecration of all the 'bizarre' ideas that we have discussed in this chapter: spacetime, energy, matter, curvature .... Clearly, these ideas were not bizarre at all!

The phenomenon just described (namely, that light cannot propagate in a straight line in a curved spacetime) is one of the most useful applications of the theory of the brilliant German physicist, so much so that it has been jokingly renamed 'the gift of Einstein to astronomy'. The reason behind this statement is easy to understand. Just like the Sun, in fact, any astronomical object can act as a *gravitational lens*, deflecting the light coming from other objects and not only distorting their appearance but also amplifying it, thus making visible those objects whose image would otherwise be too weak. This is exactly what a lens does: it concentrates a beam of parallel light rays in one point – its focus.

Exploiting these interesting deflection effects, astronomers can now obtain multiple and amplified images of galaxies at huge distances from us, in what is normally referred to as *gravitational lensing*.

As a final comment, I note that this phenomenon can also occur when the object acting as a lens is not visible because it does not emit light, although it is quite massive. In this case, the presence of the 'lensing object' can be deduced from the curvature of spacetime that it induces, and which manifests itself through the distortion of images of other distant objects. A classic example of this type of lensing is that produced by so-called *dark matter*, which does not emit light. Instead, its presence is detectable through the gravitational action it produces on a galactic scale by affecting the speed at which stars in spiral galaxies rotate around the centre, and revealing that it is present in the form of a huge spherical halo.

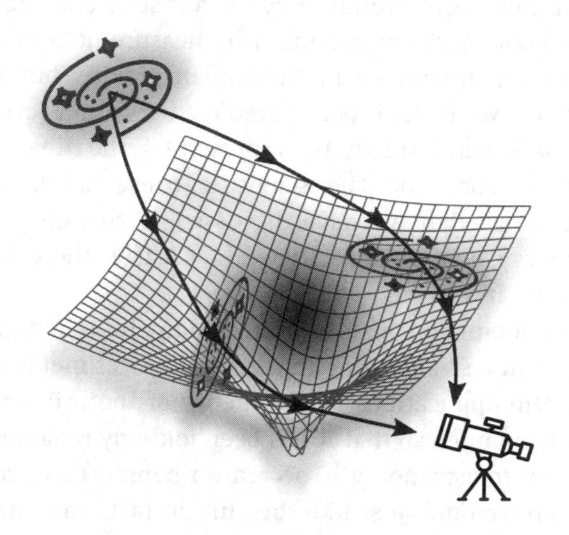

Fig. 3.10 Compact astronomical objects can act as gravitational lenses and concentrate emissions from distant sources or produce multiple images. Gravitational lenses can also be invisible, as is the case with dark matter.

Figure 3.10 illustrates schematically how the presence of invisible dark matter still produces a curvature of local spacetime, which generates multiple and distorted images of distant galaxies. In this case, by analysing how images of celestial objects are distorted, observations can be used to study the properties of dark matter, such as its mass and distribution.

But it does not end here! In Chapter 7, we will see how the deviation of a beam of light in the vicinity of a strongly curved spacetime gives rise to bizarre phenomena such as double and triple images, amplifications and absorptions. In the same chapter, I will also explain how it is possible to produce the image of a black hole, as done recently by my research group in collaboration with the Event Horizon Telescope.

4 HOW TO BEND SPACETIME

In the previous chapter, we laid the foundations needed to understand some of the fundamental concepts of Einstein's theory of gravity and, in particular, how his equations establish a strong and indissoluble link between the curvature of spacetime and the presence of mass–energy. We have also seen that curvature, when present, changes the state of motion of an object, giving rise to a dynamic that we interpret as gravity. All this leads us to two questions that can close the line of thought followed so far: 'How does spacetime actually curve? And how can we measure curvature?'.

Before providing a detailed answer, I would remind you that we have already encountered two methods by which it is possible, at least in principle, to measure curvature. The first consists of transporting a vector along a closed circuit while keeping it parallel to its initial direction; the second method, instead, involves measuring the deviation of a beam of light from a straight line. While both are valid, they are not particularly practical or easy to use. We really want a simple and effective method to get an idea of the strength of the curvature of a given spacetime, possibly simply by looking at its global properties. In particular, since the curvature is generally induced by objects with precise *macroscopic* properties (such as mass and size), it would be useful to have a method based on those properties that provide a measurement of the curvature produced by the object. In this way, we will not only learn to measure the degree of curvature, but we will also be able to *produce it!*

However, to get to what we are looking for, it is necessary to take a step back and return to what we have just discussed. We have said that the curvature radius is a length-scale associated with the radius of the sphere that approximates the curvature of spacetime at a given point. This length-scale can differ significantly, depending on whether we consider the curvature produced by an object with a very small mass (such as a grain of sand) or a very large mass (such as a supermassive galaxy). Furthermore, this length-scale can vary locally: it can be enormous if spacetime is almost flat or very small if it is strongly curved.

To obtain an indicator of curvature that works well for both a grain of sand and a galaxy, we need a *relative measurement*. In other words, a ratio between two quantities: one indicating how strong the curvature is (and therefore linked to the object's mass) and another indicating how it is distributed in space (and therefore connected to its dimensions). Mathematically, this is a very simple ratio to calculate, and the Einstein equations are extremely transparent from this point of view. In fact, they tell us that the *relative strength of spacetime curvature* can be expressed through the ratio:

$$\begin{pmatrix} \textit{relative strength of} \\ \textit{spacetime curvature} \end{pmatrix} \sim \left(\frac{G}{c^2}\frac{M}{R}\right) = \frac{(\textit{mass of an object})}{(\textit{size of an object})} \tag{4.1}$$

In this case, we must also distinguish the parts of the equation that are 'accessory', or at any rate unimportant, for the purposes of our discussion (indicated in grey above) from those that are important mathematically and physically. In Equation (4.1), in particular, the fundamental elements are the mass of the object (M) and its radius (R). Their ratio M/R in Equation (4.1) is also called *compactness*. In addition to being a very simple expression and a convenient tool to measure curvature, Equation (4.1) shows that to assess an object's ability to deform spacetime, one should not consider just one aspect (i.e., the mass or size), but both should be taken into account, as in the definition of compactness. The greyed-out parts in Equation (4.1), that is, the gravitational constant (G) and the speed of light (c), are only necessary for dividing quantities expressed in different units of measurement. For example, the mass could be in kilograms, and the radius could, for instance, be in kilometres.[1]

Hence, we could use Equation (4.1) to calculate the curvature strength where you are now as you read this book – the surface of the Earth. All it requires is to take the mass of our planet and divide it by its radius, and we find that the compactness, and thus the relative curvature is:

$$\left(\frac{G}{c^2}\frac{M_{\oplus}}{R_{\oplus}}\right) \approx \left(\frac{G}{c^2}\right)\frac{5.97\times 10^{24}\,\mathrm{kg}}{6{,}372\,\mathrm{km}} \approx 7\times 10^{-10} = 0.0000000007 \tag{4.2}$$

I think it is clear to everyone that a number like 0.0000000007 (a seven preceded by nine zeros) is extremely small. In other words, this simple calculation tells us that on the surface of the Earth, the relative curvature is essentially zero, so the spacetime is *locally flat*.[2]

However simple, this estimate is enlightening: it provides a mathematical ground for concepts that so far we have discussed only at a conceptual and intuitive level. In particular, it shows us why our perception of reality is firmly anchored in a flat spacetime and is, therefore, very far from a perception of reality that we might acquire in a curved spacetime. In other words, the difficulties we have in embracing and understanding what happens in a curved spacetime arise from it being so far removed from our daily experience of an essentially flat spacetime.

To find an analogy, think how difficult it would be for someone who has always lived in Polynesia to imagine what it means to wake up in a snowstorm. It would be difficult simply because they have only an abstract idea of that scenario. Likewise, on Earth, we live in a locally flat spacetime, which may explain why we are so surprised and intrigued by the thought that light can propagate along trajectories other than a straight line: it is a physical process of which we have no direct experience. It is inconceivable for us, at least on an intuitive level, to think we could see what lies behind a particular object if we could only capture the rays of light deflected by a curved spacetime, as shown in Figure 3.8.

Furthermore, the estimate in Equation (4.2) also explains why (with due exceptions) the theory of general relativity is not really necessary to study physics on our planet. Indeed, this is so much so that the large majority of the experiments carried out in laboratories worldwide do not take into account its corrections. For example, when studying collisions of elementary particles at the international accelerator CERN in Geneva, the interpretation of the experiments is carried out within the theory of special relativity, that is, assuming a *perfectly* flat spacetime. This is not because spacetime in Switzerland is *perfectly* flat (indeed, Equation (4.2) says it is not), but because the corrections to account for the local curvature are so small and complex that one can easily do without them.

In short, the theory of Newtonian gravitation is more than sufficient to study and calculate the effects of gravity on Earth. In fact, for most of what matters on our planet, the difference between the results it offers and those derived from Einstein's general relativity is small enough to be reasonably neglected. Therefore, engineers all over the world can continue to rely on Newton's gravity without any problem and can build the tallest of skyscrapers.

Curving Time?

While it is true that Newton's theory of gravity is sufficient for almost all practical purposes on Earth, there is an important exception that confirms this rule: the *global positioning system* (GPS). This navigation system allows us to orient ourselves when driving a car or strolling through a city. Without going into the details of its operation, it is sufficient to recall that the GPS uses a series of satellites orbiting the planet, which communicate continuously with receivers on the ground, such as our smartphones. The ultimate goal is to use the information gathered from at least three satellites to *triangulate* our position with a prescribed precision.

Essential in this triangulation process is the *time of flight*, that is, the time it takes for the electromagnetic signal to pass from us to the ground stations and then to the satellites. For an accurate calculation, extremely precise atomic clocks are used both on the ground and on satellites, with a deviation of only 40 nanoseconds per day. That is to say, the time measurements on the ground stations and on the satellites cannot differ by more than 40 nanoseconds over the 80 trillion nanoseconds that make up a day. We are talking about two parts in a million million. Despite the tremendous accuracy of these clocks, the GPS would not work without the corrections that take into account that clocks on satellites and those on Earth 'advance' with different rates, and these corrections are related to the theory of general relativity.

To understand why the 'ticking' of clocks depends on their position, we need to recall some of the concepts discussed in the previous pages. In particular, within general relativity, time plays a role similar to that of space. The presence of matter induces deformations in the elastic fabric of spacetime, changing the relative distances between its points, that is, between events.

Hence, just as the bowling ball affects the separation between different areas of the sheet (i.e., the distance between the points of a *spatial section* of spacetime in which events all have the same time), so the presence of matter affects the separation between different instants of time (i.e., the distance between the points of a *temporal section* of spacetime in which all events have the same position).

Indeed, if you think about it, we use clocks and watches to precisely measure the separation in time between two events. This can be done by exploiting different technologies: the small rotation of a gear to which the hand of a watch is connected (if it is analogue) or the oscillation of an electron between two precise energy states in an excited atom (if it is an atomic clock). In both cases, however, we expect the ticking rate, for example, of two identical Swiss chronograph watches, to be the same everywhere, in Zurich and New York. However, according to general relativity, this is only possible in one case: if they are stationary and within a flat spacetime. If, conversely, they are in motion, then even in a flat spacetime there will be a *time dilation*, as predicted by the theory of special relativity and verified every day in particle accelerators such as that at CERN. Furthermore, if spacetime is curved, as expected in the presence of matter, then the rate of advancement will also depend on the position: the same Swiss chronograph will mark time differently in Zurich and New York. This difference is due to Zurich being located at an altitude of 408 metres above sea level and, therefore, within a slightly weaker gravitational field than that of New York, which is only 10 metres above sea level.

In other words, according to general relativity (and its idea of an elastically deformable spacetime), it is perfectly natural that time is as flexible as space, and both can be curved. Equally natural is that the advancement of time is a function of the position, precisely because curvature generally changes from place to place, as shown in Figure 3.4. There, our sheet was almost flat at the edges, while it exhibited a very marked curvature near the bowling ball.

The change in the advancement of time in the presence of a curved spacetime also takes the name time dilation. General relativity quantifies with great precision this dilation on the basis of different degrees of curvature. We will return to this concept in Chapter 6 when we deal in more detail with the

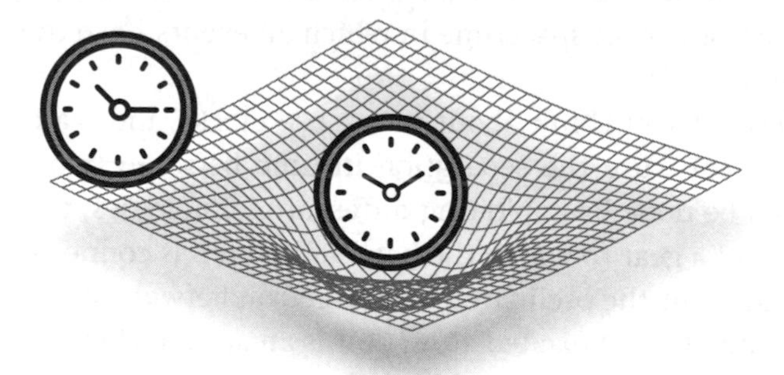

Fig. 4.1 The rate of advancement of clocks depends on the curvature of spacetime and, therefore, on their position. Those in a nearly flat area of spacetime will go faster than their counterparts located in a highly curved area.

dilation of time that occurs in the vicinity of a black hole. However, for the time being, I would like to affirm what general relativity predicts. In an area of spacetime with a marked curvature (therefore within a stronger gravitational field), time flows 'more slowly' than in an area with less curvature (therefore in a weaker gravitational field). This is shown schematically in Figure 4.1, which illustrates how the rate of advancement of a clock in a nearly flat area of spacetime will be faster than that of a clock located in a highly curved area.

To return to the GPS, this means that the clocks in the stations on Earth, where the curvature and gravity are greater, will accumulate a delay of about 45 microseconds (i.e., 45,000 nanoseconds) per day compared to identical clocks placed on the satellites in orbit at 20,000 kilometres from the Earth's surface, where curvature and gravity are smaller. This time difference is far greater than the maximum acceptable for the correct functioning of the GPS, that is, 40 nanoseconds per day. Therefore, without an appropriate calibration of the clocks, the GPS would provide completely incorrect information on our position.

I realise that all this may sound incredible or even absurd: I, too, have to summon up all my imagination to accept a phenomenon so alien to our perception of reality. However, luckily for us, what I have illustrated to you no longer represents the abstruse consequence of a bizarre theory: it is a precise prediction that today we can verify experimentally! Indeed, as science-fiction-like as it may sound, we are now able to accurately measure the difference in the ticking rate of atomic clocks that are simply placed on different floors of an ordinary building. The ability to perform such measurements has allowed us to confirm what Einstein predicted more than a hundred years ago: a clock placed on the first floor of a building, and therefore within a stronger gravitational field, will lag behind its identical counterpart on the second floor, where the field is weaker. As you can easily imagine, the difference is minuscule! So, don't try using it as an excuse for being late for an appointment – just because your watch was in a place with a high spacetime curvature . . . .

In Search of Strong Curvature

Let us now return to the estimate of curvature given by Equation (4.1). We have established that curvature is almost zero on our planet, but we can still go and 'look for' a more severe curvature in our neighbourhood, say, within the solar system. So, let us imagine building a device to measure relative curvature, and hopping from planet to planet to see how the measured values change. What would we discover? First of all, in interplanetary space, where there is very little matter, the curvature is even smaller than on Earth, and therefore spacetime is even flatter there! Curvature would become perceptible only when we get closer to a planet, but it would still have values comparable to those we have seen for the Earth. Then, as we continue our journey within the solar system, the device would reveal that the maximum value is actually reached on the surface of our star.

To know the compactness of the Sun, we just have to use Equation (4.2), substituting its mass and its radius for those of the Earth.

$$\left(\frac{G}{c^2}\frac{M_\odot}{R_\odot}\right) \approx \left(\frac{G}{c^2}\right)\frac{1.98 \times 10^{30}\text{kg}}{695,000\ \text{km}} \approx 2 \times 10^{-6} = 0.000002 \tag{4.3}$$

As we can see, the Sun's compactness is about three thousand times larger than that of the Earth and equal to 0.000002, or two parts in a million. As a result, the curvature of spacetime near our star is greater than that which we experience on our planet. This is because the mass of the Sun is three hundred thousand times that of the Earth, but its radius is only a hundred times larger. Hence, with regard to the Sun, although the denominator in the ratio *M/R* is larger than for the Earth, the numerator increases even more. However, in this case, we still obtain a disappointingly small value of curvature, so small that if we were to represent it visually through the deformation of a sheet, it would still appear flat even to those of us with the best eyesight.

Of course, if we could directly experience its effects, the curvature of spacetime near the Sun would still be sufficient to produce perceptible variations in our daily experience. In fact, the gravitational field would be about 28 times larger there; for example, assuming that on Earth you weigh about 70 kilograms, the same scale hypothetically placed on the surface of the Sun would tell you that now, weighing 2,000 kilos, you are terribly 'overweight'! In reality, your body mass would not have changed at all – you could always make up for this by weighing yourself on the Moon, thus discovering you have 'lost weight', as the scale there would measure a weight of only 12 kilos. Similarly, on the Sun's surface, the differences between clocks placed in configurations similar to those of the GPS would be of the order of hundredths of a second per day. This is an amount of time that we all have a precise idea of (think how much we use it in racing sports).

By calculating the compactness as given by Equation (4.1), we can easily determine the relative curvature produced by any object, which can lead us to surprising and otherwise unexpected conclusions. Let us take, for example, two very different objects: a star like the Sun and a galaxy like our own Milky Way. Well, the first has a mass of 1,000 billion billion billion kilograms (or, more simply, 10^{30} kilograms): a value that in astronomy is used as a unit of reference and is called the *solar mass*. The second object has a mass equal to about 1,000 billion solar masses, a number that corresponds more or less to the number of stars it contains. Such information could lead us to conclude that it produces a proportionately more severe curvature since the Milky Way has a mass considerably greater than that of the Sun. In reality, this is not the case; it is enough to compare the compactness of the two objects to realise this and understand why. In particular, the radius of the Sun is of the order of 700,000 kilometres, while that of our galaxy is approximately equal to 16,000 *parsecs* (which is about 53,000 light years, that is,

10^{17} kilometres).[3] Now, if we compute the compactness *M/R* of the Milky Way, we would realise that it is comparable to that of the Sun, namely, of the order of some parts in a million (to be precise, $M/R \simeq 1.80 \times 10^{-6}$). The reason is obvious: our galaxy is considerably more massive but also much more extended and diluted, so its compactness does not differ much from that of the Sun. In short, the curvature produced by these two radically different objects, and hence the curvature radius, is almost the same. So, if we were aboard a spacecraft and approached either one from very far away (i.e., if our spacecraft acted as what is technically called a *test particle*), we would experience essentially the same gravitational field and would have to change our trajectory using the same amount of fuel.

This example warns us of a common mistake made when thinking of curvature. In particular, when considering the curvature produced by an object, it is misleading to focus only on its mass or its size, as both quantities provide us with only a partial view. On the contrary, the ratio between these two quantities informs us how much mass *M* is confined in a certain region of spacetime of size *R* to produce a relative curvature proportional to *M/R*.

Finally, the estimates for the compactness of the Earth (Equation (4.2)) and of the Sun (Equation (4.3)) can help us to learn an important lesson. In both cases, extremely small numbers correspond to very modest curvatures and generate equally small changes in the dynamics of objects that move near these bodies. In other words, the compactness of the Earth and that of the Sun clearly tells us that although spacetime is in principle *elastic*, it is nevertheless very *rigid*. Basically, the sheet in our example is quite hard to bend!

This conclusion appears almost disarming as it would seem to indicate that Einstein's theory of general relativity is fundamentally useless. Indeed, the theory focuses on what happens when the curvature is large, but all our estimates tell us the curvature is almost always zero or so small that it can be neglected, as in the experiments at CERN.

A Theory Nobody Liked …

What has been illustrated so far, particularly the consideration that spacetime is very hard to bend, gives us a better understanding of the historical development and the early impact of general relativity within modern physics. It helps us to appreciate why no one really liked this theory, except for the few who understood it in detail and were fascinated by it. The latter includes the aforementioned Sir Arthur Eddington, an arduous defender of the theory, who introduced it to the British scientific community. A community that was not very sympathetic to the ideas of a German scientist (don't forget these were the years of the First World War and the period immediately thereafter).

When proposed, numerous scientists (even famous ones) strongly opposed the theory of general relativity. This attitude towards Einstein's innovative ideas – a rejection that began long before, with the publication of the theory of special relativity – had different roots, some even very deep. To begin with, the mathematical aspects of the theory were so complex that only few were able to follow them in detail and understand their consequences. An illuminating anecdote brings us back to Eddington's presentation of the results of measurements of the deflection of light during the total solar eclipse of 1919. It is said that a reporter asked Eddington if it were true that only three people in the world were able to understand general relativity. To which he jokingly replied: 'Oh, really? And who would be the third?'.

The attitude of rejection towards Einstein's new theory was largely due to the fact that it proposed a radically different view of gravity, in which purely mathematical and geometric considerations (namely, spacetime curvature) could actually explain what was perceived as a purely physical interaction.

In addition, the main character of this new theory (curvature) is always essentially zero, at least as far as it was experimentally measurable in those times, so it was unclear precisely what impact it could have in practice. Finally, to make the picture even more discouraging, the theory implied apparently paradoxical or even absurd consequences, such as time dilation or length contraction.

Many of these objections, and the openly hostile attitude of numerous scholars, were overcome when two of the theory's predictions, namely, the bending of light and Mercury's perihelion precession, were confirmed experimentally. However, these successes only changed the 'formal' attitude of the scientific community towards general relativity, but not the factual one. Many physicists continued to consider it (and to some extent continue to do so today) as a theory which, although not incorrect, remained of little use for practical purposes, applicable to physical conditions of curvature very different from those typically encountered in nature.

Thus, for decades, general relativity was a theory of interest primarily for mathematicians, who appreciated its rich structure and geometric elegance. On the other hand, physicists remained perplexed by the bizarre predictions and somewhat disarmed by the complexities of the equations, which often remained unsolvable, except for a few simple and idealised cases. Furthermore, given the physical conditions for which it was relevant (that were very distant from conditions usually encountered in labs and experiments), general relativity was accepted not so much because it was correct and important but rather because it was fundamentally irrelevant. In other words, Newton's theory of gravitation, however inaccurate, was simpler to use and remained accurate enough to be the only theory needed for all practical purposes.

For decades, this attitude of scepticism condemned general relativity to a state of effective neglect, denying it the development it deserved and which it would only come to know many years later. Indeed, a newly formulated theory can only offer a vague idea of the conceptual and experimental implications it can express when fully developed. To explore such implications in detail, the collective effort of the entire scientific community is necessary, which (often in no particular order) delves into the dark meanderings of the theory and gradually illuminates and reveals its various aspects. Unfortunately, none of this happened for the theory of general relativity. Between 1916 (the year of its publication) and the beginning of the 1960s, it remained the prerogative of a small group of interested physicists and mathematicians.

The turning point, and what could be considered as the 'Renaissance' of general relativity, came when new and surprising astronomical observations (in particular taken by satellites sensitive to X-ray radiation) began to provide overwhelming evidence concerning the existence of compact celestial objects whose phenomenology was inexplicable within a Newtonian gravitational framework. The incontrovertible evidence that extremely compact objects existed and were capable of releasing enormous amounts of energy forced the scientific community to reconsider the 'bizarre solutions' envisaged by general relativity almost 50 years earlier.

Einstein's revenge came when nature itself showed us that, although spacetime is nearly flat within the solar system, there are portions of the universe where it is not so at all. Moreover, astronomical observations have revealed regions of the cosmos where nature is able to overcome the enormous rigidity of spacetime, producing high curvatures that can even become extreme, as in the case of black holes. Under similar conditions, general relativity provides a mathematically adequate framework and becomes the only theory capable of explaining – with enormous success – the functioning of the physical world.

In Chapters 5 and 6, we will consider two of the most striking examples of extreme curvature nature can produce, namely neutron stars and black holes. Now, however, a final consideration awaits us.

A *Gedankenexperiment*

Before closing this chapter, I would like to invite you to perform what Einstein liked to call a *Gedankenexperiment*, that is, a 'thought experiment' conducted through imagination alone. I recall that, in general, an experiment is aimed at ascertaining whether a certain phenomenon occurs as predicted by the theory under very precise and reproducible conditions. To carry out the experiment typically requires a laboratory, an accelerator or instruments such as a satellite or telescope. In this sense, a thought experiment is something decidedly anomalous; in fact, it does not require any experimental apparatus. Instead, it is conducted by engaging only our minds on the basis of logical and physical considerations that allow us to obtain 'virtual' results that would be very difficult to produce in a real laboratory. In other words, a thought experiment is another way to find answers to questions like: 'What would happen if...?'. Therefore, all we need to conduct a thought experiment is a good dose of imagination and an excellent knowledge of physics. If you provide the first component, I will help with the second so that together we can carry out a thought experiment to investigate the limits of gravity.

So, let us start from Equation (4.1) to measure the compactness of an object. We said that the higher its value, the higher the relative rate of curvature. Since the compactness is given by the ratio between the mass, M, and the size of an object, R, we can increase the value of the ratio M/R by keeping the numerator constant and progressively reducing the denominator. From a physical point of view, we can do this in practice by taking an object, for example, a star like the Sun, and compressing it progressively to reduce its radius. Imagine a huge press that 'squeezes' the Sun more and more and reduces its radius following our wishes – I admit you need some imagination here, but this is what thought experiments are all about! The result of the process is shown schematically in Figure 4.2, and we can analyse it together.

More specifically, the panel on the left reports the spacetime curvature when we have reduced the radius of the Sun from its initial 700,000 kilometres to about 15, that is, the size of a moderately large city. Thus, the compactness M/R has increased dramatically, from the order of two-millionths of the estimate given in Equation (4.3) to 0.1, that is, to about 10%. Here, we can see that the deformation of the surface is visible to the naked eye, even without the aid of a greyscale (in all of the three squares of Figure 4.2, the 'sheet' has a length of 60 kilometres). In particular, note that the surface undergoes almost no deformation in the

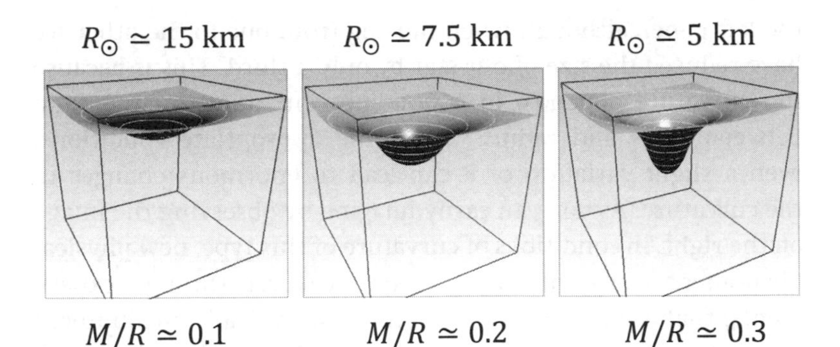

Fig. 4.2 A thought experiment on the curvature of spacetime produced by the Sun when its radius is reduced to the value indicated above each square, but its mass remains unchanged. The relative M/R compactness is indicated below each image.

areas farthest from the centre, that is, at the corners, indicating that spacetime is still essentially flat in those regions. In the centre, however, the curvature is clearly visible and has the smooth shape of a bowl.

The panel in the middle of Figure 4.2, on the other hand, shows what would happen if we continued our shrinking process and further shrunk the Sun to obtain a radius of about 7.5 kilometres, thus halving its size when compared to that in the previous box. In this case, the compactness *M*/*R* would increase to 0.2, that is to say, to around 20%. Under these conditions, the curvature becomes even more marked, and the 'bowl' produced in the sheet becomes much deeper. We can end our thought experiment with the panel on the right in Figure 4.2, where the Sun has been compressed to a radius of only 5 kilometres. In this case, the compactness *M*/*R* is about 0.3, or 30%. The curvature is very prominent under these conditions, and our sheet, which we had found to be quite stiff, has now been deformed enormously. What was a simple bowl in the left panel has become very steep-sided, resembling something that is often referred to as a 'throat'.

It is interesting to note that the degree of curvature in the right panel of Figure 4.2 is much more severe than that in the left panel, although when moving from one to the other we have reduced the size of our star by only a third. This is because the curvature depends, in a non-linear manner, on the ratio between mass and radius. Under the appropriate conditions, even a slight variation of R can lead to enormous changes in the curvature. As you can easily imagine by observing the image on the right, in conditions of curvature of this type, new physical phenomena can occur (and can be measured through experiments) that are very different from those typically encountered in a flat spacetime. We will go on to explore and explain such phenomena in due course.

I would like to close this chapter with a few questions that I hope will fuel your curiosity. We have seen what happens to the curvature of spacetime produced by the Sun when the star is reduced to a radius of just five kilometres. However, we could ask ourselves: 'Is there a limit to how much we can compress an object like the Sun? If so, what is this limit?'. And, to continue along this line of thought: 'If this limit does exist, can it be overcome? And what would happen if it could be?'.

We will find answers to these questions on the next leg of our journey.

5 NEUTRON STARS: WONDERS OF PHYSICS

In Chapters 3 and 4 we explored the concepts of spacetime and curvature and we saw how the latter gives rise to gravity. We also paused to reflect on how spacetime is simultaneously elastic and very rigid, verifying that the curvature on Earth is terribly small (which explains why we tend to think in terms of flat spacetime). Finally, through a thought experiment, we concluded that it is possible to obtain considerable curvatures by concentrating large quantities of mass (or energy) in very small volumes; for example, by 'shrinking' the radius of the Sun down to less than 10 kilometres. The moment has now come to discover together under which circumstances nature is able to produce what we hypothesised in our thought experiment.

However, to get there, it will be necessary to take a step back in time and return to the knowledge one had of the universe in the late 1950s and early 1960s. In those days, scientists had achieved an accurate understanding of the thermonuclear processes that take place inside stars that allow them to produce energy through the fusion of light elements, such as hydrogen and helium, into increasingly heavier elements, such as carbon, nitrogen and oxygen, up to even heavier ones, such as nickel and iron. The overall knowledge of this subject – which resulted from a fascinating synergy between nuclear physics, gravitational physics and astronomy – had made scientists highly self-confident and convinced that they were able to understand the way stellar objects functioned.

It was an extremely complex theoretical framework, and a huge amount of knowledge was necessary to analyse in detail the nuclear processes that occur in the most inaccessible depths of the stars (as well as to follow the numerous evolutionary paths they can take depending on their age and composition). However, the basic impression was that nuclear astrophysics and astronomy were finally able to reconstruct and explain the multiple scenarios that stellar astrophysics envisaged. Therefore, the widespread feeling within the scientific community was that (except for a few less-clear aspects) the big picture of stellar astrophysics was almost complete and essentially under control. All the observations carried out up to then were, in fact, perfectly compatible with the progress made in the study of elementary particles and with what was known, in those years, of modern physics. Finally, as far as gravity was concerned, Newton's time-honoured theory provided all the elements necessary to describe stellar physical phenomena, both for objects such as the Sun and for those with masses even a hundred times larger. Of course, there was also Einstein's theory of gravity that could be invoked, but most considered it more like a complex mathematical construct: of great charm, but of little effective relevance in the context of stellar astrophysics – and, believe it or not, many astrophysicists still think in this way!

In this climate of confidence and 'cognitive optimism', astronomers felt the increasingly pressing need to also conduct observations in electromagnetic bands that are naturally inaccessible from Earth, to determine whether they revealed sources other than those present in the optical band. To do this, it was necessary to build detectors mounted on rockets and launched beyond the atmosphere or, even better, to use satellites, which were becoming more accessible at a technological level and were also useful for purposes that were not merely scientific.

Such a massive project required enormous economic investment, but the initiatives of scientists to open new observational windows and explore new horizons did not fall on deaf ears (in contrast to what almost always does happen, unfortunately). Let us not forget that the ideological and technological confrontation between the two blocs (the United States and the Soviet Union) was very tense in those years, and the Cold War climate was also beginning to have repercussions on scientific progress.

To politicians, the idea of sending a rocket or satellite capable of intercepting X-ray emission into orbit did not seem so bad. In addition to identifying the emissions of cosmic origin, a similar instrument could potentially also identify those emissions coming from the Earth, perhaps caused by nuclear tests of the adverse bloc. For once, it was not so difficult to convey the excitement of science to politics, and so there was an abundant allocation of resources in various countries for the advent of what is now called *X-ray astronomy*.

At the centre of that new stage – on which incredible celestial bodies characterised by spectacular emissions would perform in the following years – two sources, in particular, stood out that later became true icons of X-ray astronomy. Both are linked to cases of exceptional spacetime curvature and, therefore, extreme gravity and deserve a special place among the stages of our journey.

X-Ray Astronomy and the Demolition of Certainties

In order to understand why it was necessary in the 1960s to go into space to extend our observational window, we should recall that the spectrum of electromagnetic radiation emitted by celestial objects is quite vast; ranging from the radio band, with long wavelengths and thus very low frequencies (for example those between 10 MHz and 100 MHz), up to the gamma-ray band, where shorter wavelengths correspond to higher frequencies (such as those over 10^{19} Hz).[1] The most interesting part of the spectrum is produced between these two bands, that is, the infrared; the visible (also referred to as the optical) band; the ultraviolet; and finally, the X-ray band.

Our atmosphere (the very thin but essential gas skin that covers the planet) lets much of the radiation from celestial bodies pass through it, and that is why we can witness the nightly spectacle of the starry sky or 'listen' to the universe in the radio band 24 hours a day. However, luckily for us, the atmosphere also absorbs almost all of the high-frequency radiation that would be harmful to living beings, such as ultraviolet, X-rays and, of course, gamma rays. Between the end of the 1950s and the beginning of the 1960s, there was a great technological and experimental effort to escape the constraints imposed by the atmosphere and to perform the first observations from astronomical sources in the X-ray band.

First, by using large balloons and later employing satellites, astronomers launched a campaign of observations that gave rise to the so-called *X-ray astronomy*. Obviously, and perhaps naively, it was assumed that the most intense source of X-ray radiation among the detectable sources would be from the Sun, in particular its outermost layer (known as *the solar corona*), given that the interior of the star is opaque to the X-ray radiation, which is, therefore, trapped in its interior. However, once the instruments for the first observations were built and launched, the entire astronomical community faced an enormous surprise, something exceeding even the most optimistic hypotheses. In fact, beyond the atmosphere, several sources were detectable in the X-ray band and none of the brightest ones corresponded to the Sun! One source, in particular, was so intense – with a brightness exceeding one million times that of the Sun – that it alone was able to obliterate every other source. It was evocatively named Sco X-1 because it was the first X-ray source (hence 'X-1') within the constellation of Scorpio, or Scorpius (from which comes 'Sco'). Sco X-1 was discovered in 1962 by a research group headed by the Italian physicist Riccardo Giacconi (awarded the Nobel Prize for physics in 2002), who used a rocket with a low-energy X-ray detector on board.[2] The fact that Sco X-1 was immediately spotted and looked so bright was only partially surprising and was certainly the easiest aspect to explain. It is quite common at the launch of a new mission – with the aim of making the first measurements in an observational window never probed before – that the brightest objects are also the first to be revealed. As an analogy, imagine you have some earplugs, and you take them out in a large room full of people talking. The voices you hear first will not necessarily be those of the people closest to you, nor those discussing the most interesting things, but rather those with the loudest volumes.

What left the astronomers of the time most astonished was the comparison between the position of the source and the photographic plates of that region of the sky. They discovered that Sco X-1 was associated with a decidedly ordinary and, all in all, not very interesting little star: V818 Scorpii. You can imagine the surprise and excitement of those who first made an image in the X-ray band of that object, positioned in an essentially dark and 'trivial' corner of the sky and yet endowed with an enormously greater X-ray brightness than might have been expected. In short, those very first experiments revealed that it should absolutely not be taken for granted that a celestial object luminous in the optical band ought to be equally bright in other bands. At the same time, they suggested the existence of objects that emit radiation in the X-ray band but are practically invisible in the other bands.

Subsequently, a series of observations showed that behind the emissions of Sco X-1 there was not one single object but a binary system, that is, a system containing two stars orbiting one another, clearly indicated by the fact that the brightness of Sco X-1 varied cyclically, with a period of about 0.8 days (0.7873, to be precise). The masses of the individual components of the binary were therefore measured and turned out to be quite different from each other. In particular, the most massive star, called *primary* precisely for this reason, has a mass equal to about 1.4 times that of the Sun. The second star is decidedly less massive, and therefore called *secondary*, with approximately just 0.4 solar masses. In total, the entire system, therefore, reaches about 1.8 solar masses. Today, we know that such configurations are quite common and represent a distinct class of astronomical sources called *low-mass X-ray binaries* or, simply, LMXBs.

Given the difference in mass, the two stars also had to have different sizes. But despite what one might have expected, the primary turned out to be much smaller than the secondary (although it is still not possible to determine the radius accurately). Finally, another aspect emerged rather quickly: the true source of X-ray radiation could not be the secondary star. The latter was identifiable with V818 Scorpii, the abovementioned 'ordinary' star, with a very weak brightness in the visible band. The primary, on the other hand, was shrouded in an aura of mystery ....

One of the first questions that Sco X-1 posed to the astronomical community over those years was, therefore, linked to the origin of its X-ray brightness, which was so extraordinarily high. In fact, all previous studies had assumed that thermonuclear reactions produced the luminosity of a celestial stellar-type object. However, a star of 1.4 solar masses could not produce such intense brightness (not forgetting that Sco X-1 is a million times brighter than the Sun, but has a mass only slightly larger), especially in a band of the electromagnetic spectrum at such high energy and considering that much of the X-ray radiation produced should remain trapped in the stellar core. There was, therefore, something else behind that enormous emission, something that had nothing to do with nuclear fusion processes.

Being a binary system, it was already hypothesised from the first observations that a process of *mass transfer* must occur between the two components. As is frequently the case in similar systems, the more massive star was believed to be 'stealing' material by ripping it away from its companion, and that this matter would then fall back onto its surface in the form of a so-called *accretion disc*. In the process, the matter stolen from the secondary could heat up upon impact with the surface of the primary, emitting radiation in the X-ray band. An artistic rendering of this process is shown in Figure 5.1.

However, something was wrong with this otherwise very reasonable scenario. To produce such intense radiation, it is necessary that, in the transfer from one star to another, the matter undergoes a huge gravitational jump, much larger than it would be reasonable to expect based on the assumed dimensions of the two stars.

To better understand this phenomenon, try to think of a hydroelectric power plant or a water-mill. Both structures work according to the same physical principle: the conversion of

Fig. 5.1 Artistic impression of the accretion of matter in Sco X-1. The compact primary star strips matter from its companion, which is larger in size but less massive. The matter stolen in this way forms an accretion disc that slowly pours matter onto the neutron star. © NASA/CXC/M. Weiss. A black and white version of this figure will appear in some formats. For the colour version, please refer to the plate section.

energy from one type, the *potential-energy* form, to another, the *kinetic-energy* form. In other words, when the water is taken from an elevated position, its fall is exploited to do work: to operate the turbines that produce electricity or to move the wheel that grinds the grain and transforms it into flour. The water can fall off from hundreds of metres or from just a few metres, which clearly has an impact on the amount of work produced. More precisely, we are converting the *gravitational* potential energy into kinetic energy. The gravitational potential energy is the energy an object (in this case, the water) has by virtue of being in a gravitational field and is capable of releasing if dropped, transforming it into kinetic energy – this is the energy an object derives from being in motion at a given speed (a stationary object has zero kinetic energy).

In essence, this is the same conversion thought to occur in the case of Sco X-1: the matter subtracted from the secondary star (which represents the water in our example) would have acquired a certain amount of kinetic energy in the fall towards the primary, more massive but also more compact. Finally, at the moment of the collision between the matter and the surface, the kinetic energy would have undergone a further process. Instead of being transformed into electrical energy, as in the case of a hydroelectric power plant, it would have produced electromagnetic radiation in the X-ray band.

However, as we anticipated, something was wrong with this picture. Judging by the energy produced, the gravitational leap that the matter torn from the secondary star experienced before falling to the surface must have been enormous, much larger than one would have expected if the primary had been an ordinary star of 1.4 solar masses. Returning to the example of the hydroelectric power plant, it was as if the water was falling from a mountain almost a million times higher than one would have assumed was driving the turbine downstream. In short, the obvious discrepancy between the theoretical expectations and the observations indicated rather clearly that there was something decidedly wrong with the theoretical approach and, in particular, with the type of star that the primary could be.

A further piece of observational evidence contributed to making the scenario even more enigmatic: the X-ray source was also highly *variable*. In essence, the brightness of Sco X-1 showed significant variations over time, with differences of 10 or 20 percent on timescales of minutes or even tens of seconds, much shorter than expected for an ordinary star of 1.4 solar masses.[3]

The coup de grâce in this decidedly confused and disorienting picture came with the discovery (in 1964) of another highly luminous X-ray source, baptised as Cygnus X-1 (or simply Cyg X-1). As you can guess, it was the first source in the X-ray band at the interior of the constellation of Cygnus, the Swan.

It was once again a binary system, but the masses in question were decidedly larger: the primary was estimated at 40 solar masses, the secondary at 15 (therefore, baptised as *a high-mass X-ray binary* or HMXB). At any rate, the brightness in the X-ray band of Cyg X-1 was also very high, so much so that it was anomalous. In addition, just as for Sco X-1, the binary system

Fig. 5.2 Image in the X-ray band of Cyg X-1. The emission is linked to an accretion disc that extracts matter from a 40-solar-mass star and dumps it onto a very compact object of 15 solar masses. © NASA/Marshall Space Flight Center. A black and white version of this figure will appear in some formats. For the colour version, please refer to the plate section.

of Cyg X-1 also had a short orbital period (of about 5.6 days), and this allowed astronomers to see the emission in the optical band of the primary, a star well known to be massive and bright, but otherwise relatively standard.[4] An image of Cyg X-1 in the X-ray band is shown in Figure 5.2 and allows us to appreciate how difficult it is to have a well-resolved image in the X-ray band!

Once again, however, what made Cyg X-1 truly mysterious was its variability: it could drop below one second and even reach the order of a few milliseconds. As we have seen, such a short time of variability implied very small dimensions, well below 300 kilometres of radius. In those years, no known model could explain how such a small star could produce such an emission in the X-ray band.

In short, the astronomers of the early 1960s were faced with a difficult puzzle to solve. Observations of Sco X-1 and Cyg X-1 demolished many of the certainties that nuclear astrophysics had taken about 30 years to build. There was no doubt: the observations revealed sources of very small size (and very little on any 'astronomical scale') and, at the same time, extremely massive (or having a very large mass), which produced enormous quantities of high-energy radiation (particularly in the X-ray band); stellar astrophysics could not explain their nature and nuclear physics did not know any mechanism capable of shedding light on their enormous efficiency in energy production. At the same time, however, Sco X-1 and Cyg X-1 also represented an incredible opportunity – there is nothing more interesting for a scientist than discovering something apparently inexplicable!

It soon became apparent that there was no solution to the puzzle of the two sources unless one considered the bizarre theory of gravity proposed almost 50 years earlier by the German physicist Albert Einstein. A theory predicting the existence of extremely small objects capable of generating a high spacetime curvature.

In the rest of this chapter, we will learn what we know today about that wonder of nature that is responsible for the emission of Sco X-1.

What Remains of a Massive Star

As happens almost systematically in astrophysics, the flourishing of a myriad of theories and scenarios attempting to explain the observations of Sco X-1 and Cyg X-1 soon followed. Most of these hypotheses turned out to be far from reality, even though each was physically correct. This is a rather common phenomenon in a scientific field such as astrophysics, in which we are called upon to deduce physical functioning mechanisms through simple observations without being able to create an experimental apparatus to test this or that hypothesis. We will return to this topic at the end of Chapter 6.

The first to propose the model now believed to correctly explain the phenomena related to Sco X-1 was the Soviet astrophysicist Iosif Shklovsky who (in February 1967) concluded that all previous assumptions were unsatisfactory; each had aspects that appeared unrealistic or insufficient to account for the observations. However, he was convinced that there was a fairly simple approach to explain the phenomenology of Sco X-1: his model required the formation of an accretion disc which, starting from the secondary star, caused matter to fall onto an extremely compact object, that is, a *neutron star*.[5]

What has always surprised me about the – correct – hypothesis put forward by Shklovsky is that, at that time, no one had ever observed a neutron star! Since the 1930s, after the discovery of the neutron, the existence of a stellar object composed only of those electrically neutral elementary particles was believed to be theoretically possible, but that was all. Walter Baade and Fritz Zwicky proposed this idea in an article that proved to be one of the most forward-looking in astrophysics, discussing concepts such as a supernova explosion (which we will discuss shortly), a neutron star and the origin of cosmic rays.[6] The idea of a neutron star proposed by Baade and Zwicky was certainly not part of an organically developed theory. Indeed, it appeared literally in a footnote of the 1934 article. Yet, in 1967 that was enough to change the understanding of Sco X-1 and put aside different, fascinating but incorrect interpretations.

Shklovsky's model provided a simple explanation for what emerged from the observations, albeit at the cost of introducing an 'exotic ingredient' into his recipe. Namely, an elusive compact object whose existence and origin were plausible from an astrophysical point of view (it was known that massive stars could explode and, perhaps, leave a residue), but that no one had ever identified before. There remained, therefore, a margin of uncertainty, which collapsed at the end of that same year. In Cambridge, the British student Jocelyn Bell and her supervisor Antony Hewish identified an astronomical object capable of generating a pulsed and regular radio emission – the first discovery of a neutron star.[7]

But what is a neutron star, and how is it formed? To answer this question, we must go back to what we saw at the beginning of the chapter when I recalled that a star produces energy through a series of processes of thermonuclear fusion of elements with increasing *atomic mass*.[8] Hydrogen fuses to helium, helium to carbon, carbon to neon and so on. An essential aspect for the success of this phenomenon is the production of a certain amount of energy each time it occurs. The amount of energy released depends on the elements involved in the fusion: as if climbing a ladder, the released energy increases in step with their atomic mass. Thus, the transformation of hydrogen into helium produces less energy than the next step – that from helium to carbon. More importantly, each of these fusions releases energy in the form of heat and, for this reason, they are all called *exothermic thermonuclear reactions*.

As is inevitable, however, this scale has an upper limit, at which point it is no longer possible to produce energy. Iron represents this limit, particularly by its isotope with an atomic mass of 56 (indicated as ^{56}Fe). Consequently, the fusion of two of its atoms requires more energy than it releases because it is necessary to supply the system with enough energy to bring the atoms together and merge them. Fusions of this type, which absorb energy from the outside in the form of heat, are called *endothermic thermonuclear reactions*.

Not all stars in their lifetime reach the top of this ladder; indeed, most of them (that is, those below a dozen solar masses) stop much earlier. The reason is very simple: in order to develop, fusion reaction processes need critical temperatures and densities, below which they are not triggered. This is because, in general, the atoms that take part in these reactions have a positive electric charge. Therefore, it is necessary to reach sufficiently high temperatures and densities to allow them to overcome the repulsive push that would tend to make two atoms with the same charge move apart. On the other hand, the greater the mass of a star, the greater the temperature and density inside it. Thus, ‘light’ stars (with small mass) burn only light elements, while ‘heavy’ stars (with large mass) can burn both light and heavy elements – that is to say, atoms with a large number of protons and neutrons, such as neon and silicon. In particular, stars of the order of 20 or 30 solar masses are massive enough to reach internal temperatures sufficient to trigger reactions involving heavier elements and, therefore, can climb the ladder up to iron.

Since temperature and density increase as we go down towards the centre of the star, the heavier elements are produced in the innermost regions of the star. Conversely, the lighter ones form in the outermost areas. Over its lifetime, a massive star develops a structure similar to an onion: the outer layers are rich in light elements, while the innermost ones are rich in heavy elements. If the star is massive enough, its core will be composed of iron produced by fusing the heavier elements, such as neon, oxygen and silicon.

When all the fuel that can be turned into iron runs out, the star is at a fundamental turning point because the fusion into heavier elements will absorb energy rather than produce it. From then on, nothing will ever be the same. As you can imagine, the impossibility of generating new energy compromises the balance preserved up to that moment. For millions of years, it was precisely the nuclear fusion processes that released the energy necessary to resist the force of gravity, which otherwise would make the star implode.

At this point, it may help us to understand what happens if we resort once again to an analogy. Imagine a star to be like a hot air balloon, beautifully inflated with hot air thanks to the jet of a powerful burner. From a physical point of view, the balloon remains stretched because the pressure inside it is higher than that outside. The release of energy that occurs when the burner combusts its fuel is the source of such pressure. In a star, the pressure is instead produced from the energy released through nuclear reactions. When the internal pressure is no longer available because the burner has run out of fuel, overwhelmed by external pressure and gravity, the balloon will simply deflate. The star, on the other hand, once it has run out of fuel and cannot produce more iron, is forced to implode under its own *gravitational weight*, collapsing towards the centre.

When this occurs in a star of 20 or 30 solar masses, its core is composed entirely of iron and has a radius of about 10,000 kilometres. Unable to bear its own weight and deprived of the support of pressure, the iron core contracts quickly, free-falling towards the centre (just like a hot air balloon would). During this phase, the matter at the stellar core is further compressed, reaching extreme temperatures and, above all, densities. The implosion continues until the density in the stellar core becomes such that it 'packs' the atoms into a very small volume, squashing them against each other. A new, incredible pressure develops at this point, the origin of which is not thermodynamic (i.e., linked to the fact that the temperature must increase if the density increases) but instead quantum mechanical. The atoms present in the free-falling iron core are compressed so much that the rules of quantum mechanics (according to which, particles such as neutrons cannot be too close to each other) give rise to the so-called *degeneracy pressure*, which suddenly opposes the contraction.[9]

This new and enormous pressure is able to stop the gravitational collapse of the iron core, which has now reached a radius of only 100 kilometres and a temperature of billions of kelvins. It is not possible to contract beyond that incandescent core, whose matter is the densest and most rigid we know. In the meantime, however, the rest of the star continues to fall, 'unaware' that the contraction of the core has stopped: only the innermost part is subject to the degeneracy pressure, which creates the conditions for a spectacular shock wave. At a certain point, and not being able to proceed further, the stellar mantle 'bounces' on the surface of the iron core, generating an enormous *shock front* that travels outwards from the heart of the star at supersonic speed. Already saturated with the gravitational potential energy gained during the mantle's fall, the shock front is further 'pushed' out by the energy of the neutrinos emitted copiously from the hot and dense core. The result is both catastrophic and spectacular: the shock wave is able to cross the entire star, sweeping away (at enormous speeds) all of the matter it encounters. In a few tens of seconds, a star of 20 or 30 solar masses is pulverised by a huge explosion, namely, a *supernova*. Therefore, a supernova explosion is nothing more than the phenomenon produced by the central part of a star that bounces on hitting the hard core created by the degeneracy pressure.

As the outer layers of the star are blown away, an enormous amount of radiation in the visible band accompanies the explosion, which makes the process detectable even at very large distances. If the phenomenon occurred near us (for example, within our galaxy), it would be possible to see it with the naked eye, even in broad daylight! And that's exactly what happened in 1054, when some Chinese astronomers recorded the presence, in the daytime sky, of a new star. We now know this was a supernova in the constellation of the Crab (Cancer). These days, astronomical observations allow us to study the 'remnants' of explosions of this type in what astronomers refer to as *nebulae*. These are composed of the filamentous residue of all the stellar matter that the explosion has blown away, projecting it at enormous speed, in the order of 1,500 kilometres per second – in astronomical volumes equal to 10 light years. For comparison, the solar system has a size of only five light hours.

An very nice image of the Crab nebula can be seen in Figure 5.3, which actually shows the superposition of two images taken with different filters so as to highlight the various components of the emission from the nebula. Note how this technique allows us to clearly appreciate the presence of the numerous filaments emanating from the centre of the nebula and are the result of the supernova explosion. These are what is left of the exploded star! Right at the centre of this nebula there is a tiny neutron star that, as we will learn soon, is simply a condensate of extremes.

Supernova explosions are very common in astrophysics, and estimates show that, in our galaxy alone, the Milky Way, one supernova can be observed every 30 or 40 years or so. The last one (namely, SN1987A) became visible in 1987, over 30 years ago, and astronomers are eagerly awaiting the next one . . . it will certainly be a memorable sight, which we will observe in many different ways, as I will explain presently.

Fig. 5.3 Superposition of different images in the visible part of the spectrum of the 'Crab nebula', the remnant of the supernova explosion that took place in 1054 in the constellation of Cancer. © The Liverpool Telescope. A black and white version of this figure will appear in some formats. For the colour version, please refer to the plate section.

A Condensate of Extremes

Let us go back to Sco X-1 and to what we want to discuss most in this book: gravity. There is an object linked to one of the most extreme expressions of physics and, indeed, of gravity whose appearance accompanies the incredible fireworks produced by a supernova explosion.

Almost invisible in the centre of a nebula hides the core of the exploded star that survived its collapse, an extremely dense and compact celestial body whose properties cannot but arouse wonder. Personally, it has always struck me that the final act of the bright and hectic life of a massive star (as well as the catastrophic process that reveals its death) also marks the birth

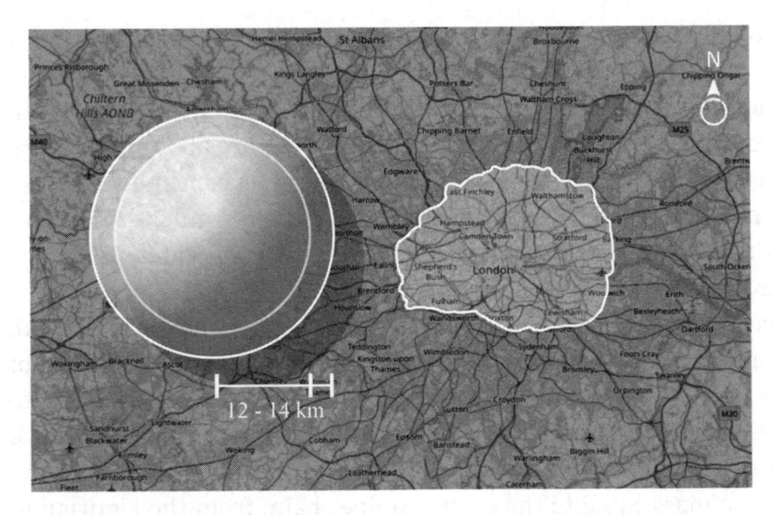

Fig. 5.4 The illustration compares the typical dimensions of a neutron star and those of the inner part of London, bounded by its north and south circular roads. © L. Weih/LR/map. OpenStreetMap contributors (CC-BY-SA 2.0)/R. K. Lazenby. A black and white version of this figure will appear in some formats. For the colour version, please refer to the plate section.

of one of physics' most fascinating notions: a neutron star. It is an object whose radius reaches about a dozen kilometres but which concentrates in itself a mass much larger than that of the Sun (examples of neutron stars with two solar masses have been observed).

Therefore, you should try to visualise a celestial body the size of the central part of a metropolitan city like London or New York, but whose mass is enormous and whose density is unimaginable for our sense of physical scales. In a schematic and explanatory way, this striking contrast is shown in Figure 5.4, which compares the typical dimensions of a neutron star (with an estimated radius of between 12 and 14 kilometres) and those of the inner part of London, bounded by its north and south circular roads. The River Thames flows through the city west to east, and further out the ring of the M25 orbital motorway may be seen . . . .

We have stressed several times during this journey that our sense of reality and physical scale is necessarily determined by what we experience on this planet. Therefore, it is difficult for us to even imagine the physical conditions that are associated with a typical neutron star. But we can still try, starting with its density (that is, how much matter you concentrate in a given volume). If we want to refer to a density we are all familiar with, we could consider water: one gram per cubic centimetre, or 1 g/cm^3. The average density of rocks is rather similar: 2.6 g/cm^3, therefore a couple of times larger than that of water. Well, the average density of a neutron star is about 10^{14} g/cm^3, while at its centre, it can even reach 10^{15} g/cm^3. In other words, we are talking about densities that are a million billion times that of water. If you want an even more direct comparison, try to imagine a single cubic centimetre of material coming from a neutron star – that is to say, as much as a sugar cube – containing a mass equal to the entire Alpine chain, from the Ligurian to the Friulian Alps. Warning. I am not referring to the weight here – although the analogy would still be correct – but to the number of atoms (actually protons, neutrons and electrons) contained in our neutron star cube. If we counted them one by one, we would discover that their number is comparable with that of the atoms of the mountain range!

Clearly, these are densities unimaginable for us, reached when all the elementary particles are compressed against each other, and the wave functions that describe them are almost overlapping.[10] To better understand this concept, try to think of an atom in normal conditions (those present in the matter that makes up you and me) as a cotton candy ball, made mostly of air and strands of sugar far from each other. In this analogy, the cotton candy represents the *electronic cloud*, namely the volume occupied by electrons (or rather, by their wave functions) in their whirling rotations around the nucleus. There is something very dense and heavy at the centre of the cotton candy, similar to one of those pellets used for fishing: the nucleus of the atom. To understand the relative dimensions and how an atom is essentially 'empty', picture the cotton candy cloud as being about 100,000 times larger than the nucleus, a tiny dot but much 'heavier' than all the rest. So, starting from this image, we can imagine, in the atoms composing a neutron star, the virtual wiping out of the cotton candy clouds (i.e., the electronic clouds) and only the tiny dots remaining at their centre. In addition, these dots are packed on top of each other, at densities more extreme than in a can of sardines. The electrons are still present in this huge clump of nuclei but are forced to move in very small spaces and cannot create the soft clouds that they would under normal conditions.

It would be nice (and very interesting) to reproduce these physical conditions of extreme density in a laboratory experiment to understand, through direct measurements, how it is possible to compress up to that point the wave functions that describe the elementary particles at the centre of the atomic nucleus. Unfortunately, as we will see better later, this is not possible; the densities that we are able to reproduce in the laboratory – even by colliding heavy nuclei with each other, launched at speeds close to that of light inside the most powerful accelerators – are much lower than those that nature manages to reach in a neutron star. This creates enormous uncertainties in our understanding of nuclear physics, the structure and composition of neutron stars. At the same time, however, it leaves us with an equally disproportionate desire to study and discover in order to find out how neutron stars create such physical conditions!

An unimaginable density is just one of the almost science-fiction aspects that characterise neutron stars. Another one that amazes me every time is their ability to rotate at very high frequencies. Indeed, we now know that some of them rotate around their axis at a frequency of almost 700 Hz, meaning that they make 700 revolutions in one second. If you can, try to visualise an object with a radius of a dozen kilometres spinning around 700 times per second. For comparison, the spin of an average washing machine reaches about 720 revolutions per minute: this means that it has a frequency of 12 Hz, that is, it performs 12 revolutions per second. In short, it rotates roughly 60 times slower than an ordinary neutron star does. In principle, we know that neutron stars could even reach 1,400 Hz. However, for some reason that we still do not fully understand, there seem to be mechanisms in nature that put a kind of brake on the maximum rotation, forcing neutron stars to rotate below the kilohertz.

Another source of wonder when thinking about neutron stars is that they are essentially perfect spheres. We discovered this because (as we will see shortly) it is possible to measure the roughness of their surface indirectly. It has emerged that the average value for the roughness is below one-millionth of the radius, at least for those stars that are not rotating very rapidly and that represent the large majority. This measure of the roughness can be taken as a measure of how large the radius variation is from one point to another. Suppose we translate this in terms of 'mountains'. In this case, it appears that, in almost all known neutron stars, the tallest surface roughness does not exceed a millimetre in height and, in many cases, it is well below one-tenth of a millimetre!

Let us try and make a comparison with the Earth, considering that Mount Everest has a height of about nine kilometres; well, the maximum surface roughness of the Earth's surface is just over one part in a thousand (1/1,500, to be precise). Therefore, although when seen from the Moon our planet appears to us as a very good approximation of a sphere, it is actually a hundred thousand times less spherical than a neutron star. Neutron stars represent, in absolute terms, the most spherical objects known to us: they far outweigh any sphere we can produce in a high-precision laboratory.

But how do we know? Obviously, we cannot go on their surface to measure their roughness with a ruler, yet we can still get fairly accurate estimates. However tiny, these 'mountains' contain enormous quantities of matter and must, therefore, generate gravitational waves (fear not, in Chapter 8, we will find out what they are and why a mountain on a rotating star should produce them). In reality, gravitational waves produced by mountains on neutron stars have not yet been detected since they are incredibly weak. However, their mere presence would produce a certain 'slowdown' in the neutron star's rotation. Since such a slowdown has never been measured, we consequently know how high, at most, such roughness can be. In other words, if on the surface of a neutron star there were mountains of a certain height, they would cause a slowdown of a certain magnitude; since this slowdown is not measured, it follows that the mountains (if any) must be of smaller height.

Let me try to recap: we have an object with a diameter of about 20 kilometres, with a mass greater than that of the entire solar system, which can rotate at the rate of 700 revolutions per second and is so spherical that its most 'conspicuous' imperfection is below a millimetre. I think even the most imaginative writers of science fiction would have a hard time imagining it. Yet this is not science fiction; it is the irrefutable, measurable and wonderful reality . . ..

But it does not end here! Another aspect of neutron stars completely detached from our common experience is their temperature. When we think of something extremely hot, we inevitably end up visualising the interior of a volcano or the surface of the Sun. Well, such environments are terribly 'cold' when compared to neutron stars. At the moment of their birth (i.e., when they are still 'proto-neutron stars') , they reach 10^{12} K, that is, one trillion kelvins: a temperature that exceeds that of the Sun by one billion times (about 6,000 kelvins) and is ten billion times that of volcanic magma (about 700 kelvins). Such heat cannot be maintained for a long time and is mitigated by the production and emission of neutrinos and antineutrinos, which subtract energy from the star, cooling it very effectively.

At the atomic level, neutrinos are produced in large quantities by converting protons into neutrons, with the additional production of an electron to conserve the electrical charge. This phenomenon has been called the *Urca process*, in 'honour' of the Casino da Urca, a casino in the homonymous neighbourhood of Rio de Janeiro. The name was chosen by the physicists George Gamow and Mário Schoenberg, who first investigated this process in 1941.[11] The reference to the casino is derived from the analogy between the production of neutrinos literally making energy 'disappear' from the proto-neutron star with a roulette table making money disappear from the pockets of the players.

As mentioned, the Urca process is indeed quite efficient. In a few seconds, the quantity of neutrinos emitted from the supernova is 10 times larger than the number of all the particles (protons, neutrons, electrons . . .) present at this moment in the Sun. In the case of the supernova SN1987A, some of these neutrinos actually reached some detectors on Earth; in particular, 12 antineutrinos (the antimatter counterpart of neutrinos) were identified by the Japanese experiment, Kamiokande, in 1987, in two different waves and about two hours before the supernova appeared in the visible band. It was a huge scientific success. At the time, I was a young university student, and I was enormously fascinated: it was the first detection of astronomical supernova neutrinos in history, with a total of 25 subatomic particles identified by experiments scattered all over the planet, and it started the field of *neutrino astronomy*.

Returning to the process of energy loss through the emission of neutrinos, this is so efficient that in a matter of seconds from its birth, the neutron star is 'cooled' to a temperature of 10^{10} kelvins (i.e., one-hundredth of the initial temperature). Within about a thousand years – a period corresponding to the age of the neutron star in the Crab nebula – it reaches approximately one million kelvins (i.e., one-millionth of the starting value). From that point onwards, the temperature of a neutron star decreases much more slowly: at the venerable age of one million years, it still has a temperature equal to 100,000 kelvins and is, therefore, a hundred times warmer than the Sun. At any rate, since the typical energy of the particles present in a neutron star is much higher because of quantum-mechanical origin, neutron stars of this age are normally referred to as 'cold' . . ..

Revealed by a Beam of Light

At this point, you may have been asking yourself the following, and the more than legitimate, question: 'How do we know about the existence of neutron stars?'. The answer takes us back in time to 1967, and to the radio observations to which everyone now attributes the merit of having allowed the precise identification of these objects. Being born in that year is a source of great pride for me, and also solves an annoying practical problem; indeed, I never remember my age, also because it seems to change all the time .... However, I remember very well that I came into the world in the year of the discovery of neutron stars which, with the help of a little arithmetic, allows me to work out how old I am.

Personal issues aside, in 1967, some observations aimed at studying very specific objects eventually led to a discovery as important as it was unexpected (something, however, that happens quite often in astronomy). A young PhD student, Jocelyn Bell, was taking measurements with a rather rudimentary radio-wave detector built by her and her supervisor, Antony Hewish. The goal was to monitor the radio emission of objects discovered a decade earlier and known as *quasars*. Today, we know that these are supermassive black holes located at the centre of galaxies, which emit enormous quantities of radiation as a result of the absorption of matter through an accretion disc.

The emission that Bell and Hewish expected to measure was almost constant over time since the variability of these sources is extremely small. However, during the observations, the PhD student also began to record a radio emission that varied with a very regular period: 1.33 seconds. In other words, while looking for a constant radio signal, she found herself observing a source that emitted radio waves with astonishing regularity and with an incredibly short period. As you will recall, a short period variability is necessarily related to a small celestial object, which was in stark contrast to the characteristics expected of quasars.

After excluding a whole series of possible terrestrial sources and confirming that the signal came from a very specific point in the sky, Bell and Hewish formulated their own thesis: they hypothesised that it was the emission in the radio band of an intense beam of radiation produced by an object rotating so fast it had to be extremely compact. To tell the truth, the two scientists did not even overlook the suggestive hypothesis that that signal was of alien origin, sent by a remote civilisation intent on getting in touch with us. To jokingly underline that possibility, Bell and Hewish baptised the periodic signal with the acronym LGM-1, where the acronym stood for *little green man*. As bizarre as it may seem, the hypothesis of an alien civilisation trying to get in touch with us earthlings is dusted off every time a radio source is identified that does not fall among those known. The latest additions to the list of these 'alien signals' are so-called *fast radio bursts* (or simply FRBs). Their origin is still shrouded in mystery, but some interesting theories have been formulated, and I have also put forward a hypothesis . . . .[12]

But let us go back to LGM-1. Since the radio emission appeared with periodic regularity and seemed to pulsate, that type of source was given the name *pulsar*, a combination of *pulsating* and *quasar* (the objects it was originally intended to observe). The discovery of that emission represented a huge turning point in our understanding of astrophysics, nuclear physics and gravitational physics. To underline its importance, in 1974, Martin Ryle and Antony Hewish received the Nobel Prize in Physics with the official reasoning: 'For their pioneering research in radio astrophysics'. Hewish was awarded the prize precisely for the discovery of pulsars, yet no recognition was extended to Jocelyn Bell, who had collected the data from which it all started. To this day, there is still no convincing explanation for this exclusion, although Bell has always claimed that the decision of the Royal Swedish Academy of Sciences was correct.

At any rate, more than 50 years after that discovery, we now know that in 1967 Bell had identified the radiation beam produced by the surface of an extremely compact star – a neutron star, in fact – with an incredibly strong magnetic field. The details of the mechanism (or mechanisms) that produce this radiation remain partly obscure. Still, the whole scientific community has generally accepted the overall scenario: if a neutron star has a strong enough magnetic field and is subject to a high enough rotation, it is able to develop an electric field sufficiently powerful to tear electrons from its surface and, in particular, from the magnetic poles. These free electrons are accelerated by the electric field and produce radiation not only in the radio band, but also in the X-ray and gamma-ray bands, emitted in two thin cones.

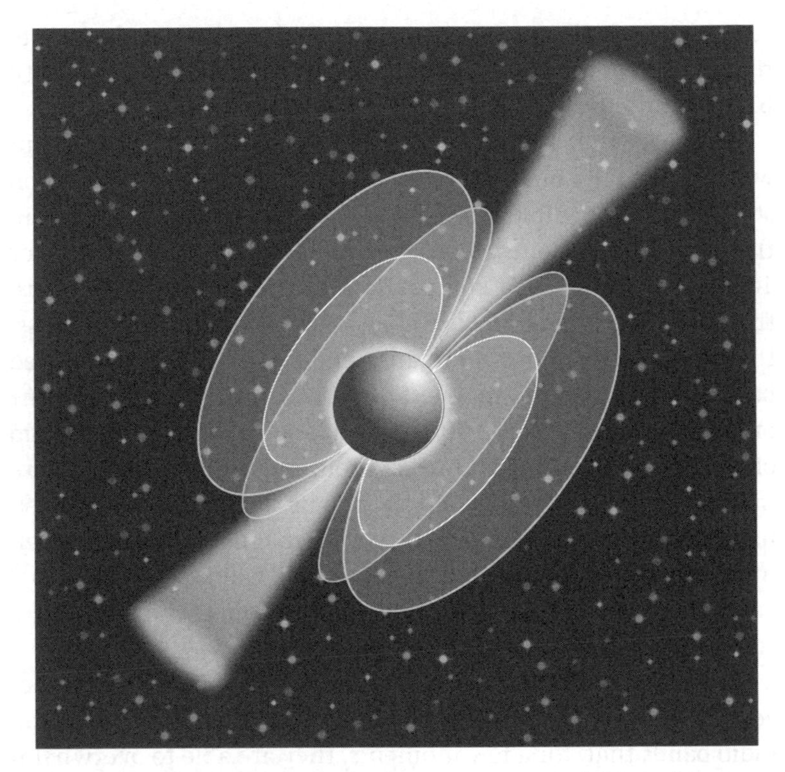

Fig. 5.5 In a pulsar, the rotation and strong magnetic field accelerate particles near the polar caps, which emit collimated radiation (especially in the radio band). The beams rotate with the star, similar to those of a lighthouse, and can 'sweep' the Earth.

The operation of a pulsar is schematically represented in Figure 5.5, which shows how, in a neutron star, the combined action of a rapid rotation and a strong magnetic field produces a thin pencil of radiation, predominantly in the radio band. The presence of a misalignment between the axis of rotation of the star and the axis of the dipole moment of the magnetic field causes the radiation beams to rotate with the neutron star, periodically sweeping different regions of the sky.

Therefore, a pulsar is a rapidly rotating neutron star, emitting a beam of light in the radio band, just as a lighthouse emits a beam of light in the visible band. So, when we on Earth are 'swept' by the beam of light emitted by the neutron star, we see a pulsed radio emission, with the pulsation period equal to that of the rotation of the star (or half of it, if we can see both beams emitted by the poles of the magnetic field).

An aspect that I have so far mentioned only in passing, but which is actually one of the fundamental properties of neutron stars, is the incredible intensity of their magnetic field: even in the most ordinary ones, it reaches values of the order of 10^{11} or 10^{12} gauss. To make a comparison with realities closer to us, the Earth's magnetic field, for instance, is about half a gauss, that of the Sun about one gauss and that of the magnetised bars (for example, the magnets we stick on the fridge) is of the order of approximately 100 gauss. In short, the magnetic field of a normal neutron star is a thousand billion times more intense than that which confronts us daily on our planet. But there are neutron stars whose magnetic field is even more intense and can reach enormous values: 10^{14} or 10^{15} gauss, or up to a thousand times more intense than that of a 'standard' pulsar.

These neutron stars take the evocative name of *magnetars* and, very rarely, can emit such powerful flashes in the X-ray and radio bands that, for a few moments, they are able to overwhelm all the other celestial objects combined. To give you an idea of the absurdly large amount of energy released by a magnetar when it emits radiation explosively, we can take the case of SGR 1806–20, which recorded one of these explosions in December 2004. Calculations show that the amount of energy emitted by this source in just one-tenth of a second exceeded that produced by the Sun over the last 100,000 years!

The cause of these phenomena is still unknown. However, it is hypothesised that sudden changes in the balance of the crust of these neutron stars (comparable to portentous earthquakes) produce massive variations in the magnetic field on the surface. In turn, this releases enormous amounts of magnetic energy, which is converted into X-ray or gamma-ray radiation. The process itself is not all that unusual, and it is observed regularly in the Sun, where it gives life to real magnetic explosions, also called *solar flares*.

One may well ask: 'How are such intense magnetic fields produced?'. In fact, the origin of the enormous magnetic field of pulsars and magnetars is still the subject of intense research and debate today. One plausible explanation – though not the only one – is that it is a 'fossil' magnetic field, that is, inherited from the original star before it collapsed. In this case, the process that comes into play is the *conservation of the magnetic field flux* (or *flux-freezing*) during the gravitational collapse that gives rise to the neutron star. The product between the section of the star itself (which is proportional to the square of its radius) and the intensity of the magnetic field should always remain the same. During the collapse, the original star goes from a radius of more than one million kilometres to one of only ten, so that the mere reduction of the section is sufficient to provide an amplification factor of $(100,000)^2 = 10,000,000,000$ times.

To understand what is happening, imagine taking a magnet, like many of us have on the refrigerator at home, and compressing it with a hydraulic press to halve its radius: in this way, its magnetic field will increase by a factor of four, that is 2^2. The same also happens in a collapsing star. The big difference is that it can start from a very weak magnetic field and develop an extremely large one due to the enormous reduction in radius caused by the collapse.

Alongside the hypothesis of the fossil origin, there is also another, namely, that the incredible magnetic field is generated after the collapse when the neutron star is still an extremely hot and fluid object, subject to large oscillations and looking for a new balance. Under these extreme conditions, complex phenomena of turbulence and a magnetic dynamo inside the proto-neutron star can amplify its magnetic field up to the values measured with the radio observations. The latter scenario, the details of which are still difficult to reproduce through simulations and theoretical estimates, could explain the exceptional magnetic fields observed in magnetars.

To complete the list of the wonderful properties (in the sense that they cannot arouse in us anything but a sense of pure wonder about neutron stars), I would like to show you one more:

they are incredibly accurate clocks. To understand how this is possible, we must remember that, by virtue of the high rotation and the ability to emit a pencil of light in an exquisitely periodic manner, pulsars represent the astronomical objects measured with the greatest accuracy: the radio astronomy of pulsars is, in fact, a very high-precision science.

To have at your disposal a signal that repeats regularly and with high frequency allows you to collect, over the years, an enormous amount of data with precise statistical properties. In this way, it is possible to reduce the systematic errors that usually plague measurements and to obtain excellent accuracy. That is why the period of some pulsars is known with an accuracy of at least 10^{-15}, that is one part in a million billion (or, in other words, with such precision that it can be expressed with a number of which we know 15 digits exactly!). For example, this is the case with the fastest rotating pulsar known to us, namely PSR J0437–4715: its period is known with an even higher precision to be equal to 0.005757451936712637 seconds.

To get an idea of how accurate this measurement is, imagine the precision needed to identify, one by one, all the cells present in a human being, which are approximately 30,000 billion (or approximately 3×10^{13}). Well, the period of PSR J0437–4715 is known with an accuracy that is a hundred times higher! If we had achieved this precision in biology, we would be able to 'clone' a human being simply by putting together the 30,000 billion cells that make it up....

The incredible precision with which we can measure the period of pulsars also tells us that we can use them as highly accurate cosmic clocks. Now, for a watch to be defined as 'good', it is essential it has a regular period, but it is equally important that its period changes in a constant and predictable way. In the case of pulsars, it changes almost imperceptibly, but it does change. So, what matters, and is lucky for us, is precisely that it varies on a basis that is regular and easy to predict.

The reason why the pulsar's rotation period must change is rather simple to understand. Just like a lighthouse (which needs electricity to function) a pulsar must tap into some form of energy to emit its beam of radiation in the radio band. This 'reserve' is represented by its kinetic energy. To be more precise, the rotation of a pulsar varies, and its period becomes longer and longer because the star loses part of its rotational kinetic energy through the emission of electromagnetic waves in the radio band (this is called pulsar 'spin down'). Recognising this mechanism represents a cornerstone in our understanding of pulsars, and the idea was first suggested by the Italian astrophysicist Franco Pacini in 1967 (once again, *annus mirabilis*!). He showed that a pulsar is similar to a rotating magnetic dipole: in short, a magnetic rod with a north pole and a south pole in rapid rotation. The rate of change of rotation is related to the amount of energy lost.[13] It is as if the emission of electromagnetic radiation works as a 'brake' for the stellar rotation, and one can even measure how strong this brake is through a combination of the rotation period and its variation over time; a measurement that takes the name of *braking index*.

However, the energy lost by the pulsar in this way is only a very small fraction of the available kinetic energy. That is why the period of another pulsar called PSR J1603–7202 – which we know to be 0.0148419520154668 seconds – increases by just 0.0000005 seconds every million years.[14] In comparison, a precision Swiss watch 'loses' about 7 seconds a day and a quartz watch about 10 a year... by now, you will have guessed how much neutron stars fascinate me. However, in order not to be 'biased', I must remind you that, with today's technology, we can build atomic clocks with even greater accuracies than those mentioned for pulsars: only 0.0000001 seconds lost every million years, five times less than a pulsar. However, it should also be said that it is possible to maintain this precision only over fairly short time spans, of the order of a few hours. Conversely, pulsars can be stable for hundreds of thousands of years. Which, in my opinion, definitely makes them superior!

I would like to conclude this section by anticipating a question that some of you may have contemplated. We have, in fact, said that a pulsar emits a collimated beam of radiation in the radio band and that to do so it requires a huge magnetic field and a sufficiently high rotation since it is the combination of these two elements that produces the enormous electric fields that can accelerate charged particles. However, we have just seen that a pulsar rotates more and more slowly, even though it will take millions of years to slow down completely. It is therefore legitimate to ask whether, at a certain point, the rotation of a pulsar can become so slow that it is no longer able to emit electromagnetic radiation. The answer is yes – a pulsar can 'turn off', that is, slow down so much that it is no longer visible as a lighthouse. When this happens, the pulsar is said to have crossed the *death line*. At that point, it will no longer be visible even though it will obviously continue to rotate and have a strong magnetic field for billions of years more. Since we only see those neutron stars showing up as 'turned-on pulsars', we believe there are many more than we have seen so far!

A Mysterious and Fascinating Interior

Another more than legitimate question to ask oneself here is: 'But what does the interior of a neutron star look like?'.

The answer, without much ado, can be summarised as follows: fundamentally, we do not know! In other words, we have only a vague idea – or, rather, many but conflicting ideas – about what the internal structure and composition of a neutron star is. The reason for this 'ignorance' is quite apparent: the physical conditions inside such an object are far from those testable and reproducible through laboratory experiments. Thus, it is objectively difficult, from a theoretical as well as an experimental point of view, to extrapolate from what is known at much lower densities (and with an enormously smaller number of particles) and reach conclusions that are valid for the density and pressure present inside a neutron star.

That said, there are aspects of their composition and internal structure that everyone agrees on. For example, it is quite clear that a neutron star is not made only of neutrons but also contains other particles within it (albeit in small quantities). The most important are certainly protons and electrons, and the latter are ultimately responsible for the production of the massive magnetic fields we have talked about. Electrons (and other light, charged particles) are, in fact, capable of producing the enormous electric currents necessary to generate them. Furthermore, despite the underlying uncertainty, it is quite clear that the structure of a neutron star must be characterised by specific areas whose thicknesses are known to us with some precision. On the other hand, the presence of layers is true also for our planet, whose physics is much simpler and whose structure we can investigate directly.

So, let us imagine that we 'dissect' a neutron star in order to understand its structure, starting from the surface and moving towards the centre. The first layer we encounter consists of a sort of atmosphere (and is technically referred to as such), a very thin skin, not more than one centimetre thick, composed of extremely heavy atoms and with a density billions of times higher than that of our atmosphere. However extreme, the properties of this atmosphere are quite clear, and its physics relatively well tested, so much so that we consider it essentially as 'known'. As paradoxical as it is, the only part of an object with a radius of a dozen kilometres that we believe we know in detail, in terms of its properties, is no more than one centimetre thick.

Moving towards the centre, below the atmosphere we will find what is referred to as the *crust*. A layer with a thickness of about one or two kilometres containing heavy ions, ions with a large atomic mass, but also extremely high-energy electrons, thus called *relativistic electrons*. It should be noted that the term 'crust' can be misleading, as it is an elastic and deformable material, somewhat similar to an extremely dense plastic substance. However, part of the matter in the crust will still have the characteristics of a lattice, that is, it will show a periodic and regular structure in which the ions are at precise distances, and the electrons are free to move between the spaces left empty. This type of lattice structure is the one usually encountered in solid objects and is responsible for their mechanical properties.

Below the crust, in a layer that could extend for six or seven kilometres, we will encounter the *outer core*; there, the density reaches thousands or tens of thousands of billions (in short, 10^{13} or 10^{14}) of grams per cubic centimetre. An enormous density but not the maximum one, which we will meet moving towards the central zone of the neutron star, its *inner core*, which is also six or seven kilometres thick.

The properties of the matter present in the inner core remain fundamentally unknown and represent an exceptional theoretical challenge, with which nuclear physicists have now been confronted for almost 40 years. You will not be surprised to know that there is no shortage of hypotheses in this regard, some more conservative, others decidedly less so. Perhaps, however, the most important and fascinating question about the physical conditions in the deepest areas of a neutron star concerns the presence of *exotic particles*, such as hyperons or even free quarks. We talked about the latter in Chapter 1: the elementary particles constituting other particles, such as neutrons and protons. In other words, it is possible that at the centre of the star, as a consequence of the very high density reached in its innermost core (whose radius does not exceed a couple of kilometres), quarks are so close to each other that they become 'free', that is, no longer confined within a neutron or proton, and thus form a *quark soup*. This hypothesis is particularly fascinating because we know that a soup of this kind must have been present in the very first moments of the universe's life (up to a hundredth of a second) and is produced, even if only for very short intervals of time, when heavy ions are collided at very high energies in experiments. The idea that this soup is instead present in a stable manner inside the neutron stars and can somehow be revealed (perhaps through the emission of gravitational waves) has the potential of opening up research areas involving scientists from all over the world, myself included.

Doubts about the internal composition of neutron stars, which we have just discussed, are generally described as an uncertainty in the so-called *equation of state*. To be precise, when physicists speak of the equation of state of a neutron star, they are referring to the ability to understand how the pressure (and therefore the internal composition) of such an object varies in the passage from the surface to the centre. The uncertainty in this regard also translates into an uncertainty – which is particularly serious from an observational point of view – about the size that a neutron star can have. Until a few years ago, the estimates of the radius of these celestial bodies ranged fairly widely: from 7 kilometres, when calculated using equations of state that predict very compact stars, and which take the name of *soft equations of state*, to 18 kilometres, when derived from those equations of state that predict comparatively larger stars and that take the name of *stiff equations of state*.

Fortunately, today we have a whole series of observations – both in the electromagnetic channel and in the gravitational-wave channel – that give us more precise indications regarding the radius of a neutron star of 1.4 solar masses (generally considered as the reference mass): it is expected, in fact, that it is between 10 and 14 kilometres. At any rate, the determination of the radius of neutron stars represents an extremely active research area, one in which I too am a part. Moreover, new estimates are suggested every time a new instrument provides new measurements or when the merger of a binary system of neutron stars is detected. It is, therefore, not excluded that the estimates given above will soon become obsolete!

Maximum Mass

The idea mentioned previously of a typical radius for a reference mass allows me to introduce an aspect that I have so far left out when presenting the properties of neutron stars, although it is the main topic of this book: the gravity produced by such a celestial body.

Also, from this point of view, a neutron star is an object with extreme characteristics. The gravity it generates is, in fact, at least a billion times stronger than that present on Earth. If, absurdly, we landed on its surface, the gravity would reduce our spaceship – and us inside it – to an almost two-dimensional object: we would be spread on a reticular surface that would extend vertically for no more than a handful of atoms.

Furthermore, since the typical compactness of a neutron star varies from $M/R = 0.17$ to $M/R = 0.25$, it is clear that we are dealing with very significant curvatures to say the least, and which give rise to very strong relativistic effects, such as time dilation or the curvature of light trajectories.

There is, however, a particular aspect in the relationship between gravity and neutron stars, which represents a direct consequence of the theory of general relativity: the existence of a *maximum mass*. To understand what this is, it is necessary to refer to the solutions of the Einstein equations that allow us to build models of neutron stars in equilibrium. Basically, after choosing an equation of state that links the pressure inside the star to the local density of matter, it is possible to solve Einstein equations and find an equilibrium model representing a neutron star. At that point, by varying the mass of the star, it is possible to obtain a 'sequence' of models, that is, the set of neutron stars that belong to a certain equation of state. Since we do not know the correct equation of state, we can construct numerous sequences of this type – I have built almost two million of them – all of which are physically plausible.

All these sequences share a common property: the stellar radius decreases as the mass increases. Thinking about it, this is a fairly anomalous behaviour in astrophysics: we are used to finding objects in which a greater mass also corresponds to a greater radius – just think of what happens to ordinary stars or planets. However, in the extreme conditions of density and pressure that characterise neutron stars, the opposite is true: as the mass grows, the radius tends to decrease.

Figure 5.6 shows a schematic plot of the relationship between mass and radius in a generic neutron star. The mass (expressed in solar masses) is placed on the vertical axis, the radius (expressed in kilometres) on the horizontal axis. The solid black line represents the set of solutions of the Einstein equations for a certain equation of state so that every single point on the black line is a configuration in stable equilibrium. As can be seen, the radius tends to decrease as the mass increases, but the latter cannot grow indefinitely. The black curve has a maximum, indicated

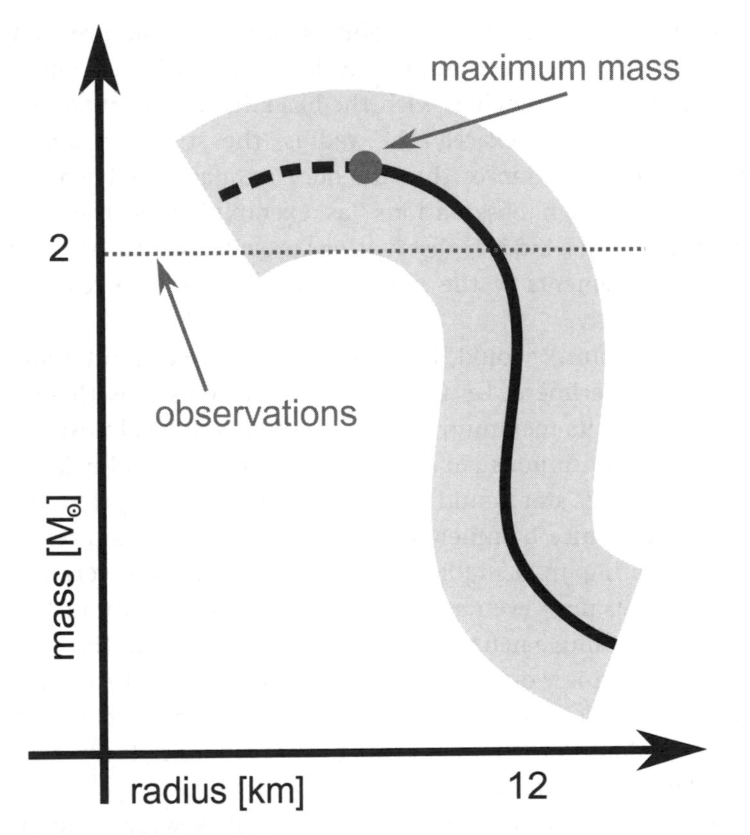

Fig. 5.6 Relation between mass and radius in a neutron star. The radius decreases as the mass increases, reaching a minimum at the maximum mass. Stars with larger masses would collapse, so they are not possible.

with a dot, beyond which the line becomes dashed: this is the so-called *unstable branch* of the equilibrium configurations. In other words, although it is also possible to find solutions to the Einstein equations in the dashed part of the line, they are however subject to a 'gravitational instability' and thus unstable. The corresponding neutron stars would collapse under the gravitational pull.

Therefore, every equation of state has a maximum mass, with a stable and an unstable branch of equilibrium solutions; the grey band surrounding the black curve thus represents the uncertainty introduced in determining the radius of a neutron star due to the ignorance about its internal structure. If the equation of state was unique and known, that band would be reduced until it coincides with the black line. Thus, it would be possible to know exactly the radius, the structure and the internal composition of the star once its mass has been fixed, perhaps through observations (as exemplified by the dotted horizontal line). Indeed, observations provide us with very accurate measurements of the mass but are not able to reveal the radius directly.

At this point, I would like to invite you to work out another thought experiment. Let us consider a neutron star with a mass very close to its maximum: in the graph, it will be almost at the top of the continuous curve, slightly to the right and below the black dot. This star would be perfectly stable; if we perturbed it by placing a nice hammer blow to its surface, it would respond by oscillating in a stable, if damped, way. However, if we increased its mass even by an infinitesimal amount, for instance, by adding a single neutron, we would take it beyond the critical point, where it would be unstable. Its mass would exceed the maximum allowed by the equation of state and would, therefore, be forced to do something very different from simply oscillating: it would collapse into itself.

A maximum mass limit is not foreseen in Newton's theory of gravity, according to which it is always possible to add a little matter to a star. Likewise, Newtonian physics does not contemplate the gravitational collapse of a neutron star with 'excessive' mass; it is the result of the existence of additional solutions to those of neutron stars, and we will discuss this in the next chapter. At the moment, we do not know the value of this maximum mass. However, there is a pulsar – PSR J0348+0432 – whose mass has been measured with great precision and found to be 2.01 ± 0.04 solar masses. Since a neutron star with such a high mass exists in nature, it follows that the maximum mass must necessarily be higher (just as shown in Figure 5.6).

Furthermore, the recent observation of the merger of a binary neutron star system – GW170817 – has provided data for new hypotheses about this limit. Currently, many scientists – myself included (as I argued in an article that appeared a couple of years ago[15]) – think that the maximum mass likely falls below 2.33 solar masses. In other words, the phantom maximum mass of a neutron star is now reasonably placed between 2.0 and 2.3 solar masses. Unfortunately, although it is a fairly narrow range, it still does not allow us to have a precise idea of the structure and internal composition of a neutron star. However, the observations in the years to come will provide us with many other indications, which will help us to obtain tighter constraints on its value.

At the conclusion of this digression on the maximum mass, and as an appetiser for what we will go on to discuss, let me anticipate that to understand what happens to a neutron star with a mass larger than the maximum one, and thus doomed to collapse because gravitationally *unstable*, we need to understand the final state that is reached through the collapse. That is, the state of a black hole, to which Chapter 6 is dedicated.

Who's Afraid of a Neutron Star?

Over the past 50 years, neutron stars, and pulsars in particular, have been the subject of intense and uninterrupted observation campaigns. For this reason, we now know of more than 2,800 pulsars, and all of the related information is freely available on the internet through the catalogue compiled and updated by the Australian Telescope National Facility (or, more simply, ATNF). However, the neutron stars observed so far represent only the tip of the iceberg; in the Milky Way alone, their estimated population varies from two hundred thousand to one million. Most of these pulsars are isolated, that is, they are not associated with other stars, but there is also a considerable fraction of them within binary systems, that is, where an ordinary star accompanies the neutron star.

This is precisely the case with Sco X-1 – the starting point of this chapter – where the neutron star strips matter from the secondary and makes it fall onto itself in the form of an accretion disc. When the matter stolen from the companion falls on the neutron star, the huge gravitational leap it experiences (remember hydroelectric power plants?) causes it to heat up, reaching temperatures of one million kelvins, thus producing the X-ray emission that allowed it to be revealed.

However, binary systems of this type can reach even higher temperatures! Thinking about it, the matter deposited on the surface of the neutron star must also go somewhere. As far as we know today, it accumulates in a layer just a few centimetres thick (let us not forget that the gravity on the surface of a neutron star is enormous). Since the ordinary star's outer parts are composed of light elements such as hydrogen and helium, the transfer of matter from the ordinary star to the neutron star only deposits matter with low atomic mass (hydrogen or helium), which is therefore highly 'combustible' via thermonuclear fusion. Thus, as the stripped matter accumulates and heats up on the surface of the neutron star, the right conditions of density and temperature can be reached in order for a thermonuclear fusion reaction to be triggered in that thin skin, which literally destroys the accumulated layer, causing it to lift off the surface into a dazzling sphere of light.

This massive nuclear explosion, involving a considerable portion of the entire surface of the neutron star, releases an amount of energy equal to nearly one billion billion atomic bombs. Furthermore, this huge release of energy occurs on a timescale ranging from 1 to 10 seconds. It gives life to the process that astronomers refer to as an *X-ray burst*, which characterises, in a manner that is more or less regular, the emission of sources in low-mass X-ray binary systems (LMXBs). Neutron stars have not only been observed in LMXBs, they are also present in other configurations and, in particular, in binary systems composed of two neutron stars. A special subclass of these systems is then represented by those binary systems in which both neutron stars are also pulsars and are called, for this reason, *binary pulsars*. We will return to them in Chapter 8, when we discover that they represent extremely intense sources of gravitational waves and very rich laboratories in which to study the properties of gravity within very strong fields.

At this point, however, I would like to conclude this chapter with a logical question, one that some of you may have already asked given the extreme characteristics – in every sense – of neutron stars: 'Should we be afraid of these celestial objects?'.

Just like any natural phenomenon of great beauty and equal power, such as an erupting volcano, it is also advisable to observe neutron stars from a 'safe distance'. This is because their properties can prove to be decidedly dangerous for us fragile human beings. However, the closest neutron star to us – namely RX J1856.5–3754 – is 500 light years away (remember that the closest ordinary star, apart from the Sun, is only 4.24 light years away from us: Proxima Centauri). In addition, neutron stars are not static and indeed normally move at relatively high speeds (between 200 and 400 kilometres per second), which they acquire at birth through the supernova explosion. The emission of neutrinos in the first milliseconds of the protostar's life is highly *anisotropic* (that is, it has different intensities in different directions), and this causes the protostar to experience a 'recoil' in a given direction, in which it will continue to wander for the rest of its existence.

Now, if a neutron star like RX J1856.5–3754 were to cross the solar system, even without colliding with any of its planets (remember it is a tiny object that, if compared to the scale of the other celestial bodies present in the solar system, would be essentially a dot, slightly larger than an asteroid), it would create 'considerable' gravitational disruption due to its imposing mass, certainly greater than that of the Sun. It would break some of the orbits of our solar system, and if it passed close to one of the planets, for example, Jupiter (the largest), it would simply destroy it thanks to the enormous tidal forces that would develop.

On the other hand, if a magnetar exploded near us, flooding us with very high-energy radiation, this would be sufficient to erase all life on Earth within seconds. This would be the case even for areas of the planet in the shadow of the X-ray emission, which would instead be flooded by the radiation reflected from other planets or the Moon. In other words, I think we can all agree that it would be advisable to avoid too close an encounter with a neutron star, especially if it is emitting large amounts of radiation.

That said, we have very little to worry about: the magnetars within our galaxy are known, and all very distant from us. And even if pulsar RX J1856.5–3754 was directed exactly towards Earth at the maximum speed allowed by a neutron star, it still would not arrive here for six million years. So, in essence, at least for the time being, we have nothing to fear from neutron stars and, instead, we can enjoy the wonders they reveal to us . . ..

6 BLACK HOLES: CHAMPIONS OF CURVATURE

We saw in the previous chapter how the advent of X-ray astronomy at the end of the 1960s left astrophysicists 'bewildered'. The observations obtained by X-ray satellites revealed the existence of never-before-seen objects, with very large masses but extremely small in size. Sources such as Sco X-1 and Cyg X-1 represented inexplicable surprises within astrophysics and called into question all that was understood of modern physics at that time. To make matters worse, a plausible explanation by way of a hypothetical 'neutron star' could be postulated for Sco X-1, but the phenomenology of Cyg X-1 represented a far more puzzling enigma.

With a mass of about 15 solar masses, it certainly was not possible to invoke the presence of a neutron star for the observations of Cyg X-1. Furthermore, the variability of the X-ray emission from Cyg X-1 (usually around one second, but potentially dropping to just a few milliseconds) suggested the presence of a much more compact object, with a size in the order of only a few tens of kilometres. Today, we know that Cyg X-1 is a fairly ordinary representative of a whole class of celestial objects exhibiting the same phenomenology: the high-mass X-ray binaries (HMXBs). Again, these are binary systems in which there is a transfer of mass from an ordinary star to another much, much more compact object. As with Sco X-1, the transfer of matter in the case of Cyg X-1 also occurs via an accretion disc through which matter falls onto the compact object, heats up and emits very intense radiation in the X-ray band. The most relevant difference between Cyg X-1 and Sco X-1 is that, in the case of

Cyg X-1, the ordinary star was the object with the largest mass, while the other more compact member of the binary is what today we know to be . . . a *black hole!*

These celestial bodies have become an integral part of the collective imagination and general knowledge. In fact, so much so that a shrewd (and amusing) answer to the simple question, 'What is a black hole?', came from my little daughter Anna, who was not even old enough to go to school at the time. Yet, as children often do, she addressed the question with the superiority typical of someone who thinks they know everything about life: 'Well, it's obvious! It's simply a hole that can't be filled . . .'. This answer is certainly not wrong – actually, it is perfectly true because normally a hole can be filled. However, it remains a purely qualitative description, which unfortunately does explain what black holes are. Since it is not uncommon that information found on black holes is confusing or even misleading, it may be helpful to put some order to what you may already know.

Thus, to remove some of the urban myths and most superficial descriptions of black holes, I will endeavour to provide a clear and gradually more detailed explanation of what they are and the properties they possess. I will begin with their genesis and end with a discussion of the difficulties that scientists still have in managing a concept that is fundamentally difficult to accept.

An Utterly Bizarre Solution

A simple yet rigorous approach to explain black holes could come from their definition and, in particular, from the fascinating, albeit tragic, genesis of what turned out to be the first solution to the Einstein equations, found by Karl Schwarzschild (1873–1916), a physicist from Frankfurt am Main.

I recall that in November 1915, in the proceedings of the then Royal Academy of Sciences in Berlin, Albert Einstein published his field equations, the foundation of his theory of general relativity. At that time, Karl Schwarzschild was already a renowned astronomer; indeed, he was the director of the Astrophysical Observatory in Potsdam, a few kilometres from

Berlin. Despite holding this prestigious role, like many German Jews of his generation, Schwarzschild felt the need to express his loyalty to Germany and the sense of belonging that bound him to German society. Consequently, at the outbreak of the First World War in 1914, he enlisted as a volunteer and was sent to fight. Although already aged 40, he was sent to the western front, and then to the eastern front. Unfortunately, due to the harsh conditions encountered in Russia, he became seriously ill, and his health quickly deteriorated. He died in 1916 but, in the last months of his life, he managed to find the first solution to the Einstein equations.

Unaware of the impact it would have, Schwarzschild gave no name to his solution. In 1967 (another significant event that year...) the American physicist John Archibald Wheeler used the descriptive term 'black hole'.[1] Although Schwarzschild's findings represent the simplest solution of the field equations of general relativity, it is only possible for me to cover this topic in the final part of my course on general relativity in Frankfurt, after spending months on introducing all the mathematics necessary for its derivation. Let me remind you of the three important hypotheses under which this solution can be found:

(1) the presence of a spherical symmetry;
(2) the independence from time, that is, the solution is static; and
(3) the absence of matter, that is, the solution is devoid of matter and therefore in what is called a 'vacuum'.

We now know that Schwarzschild's is also the *only* solution to the Einstein equations that satisfies these conditions (this is the thesis of the famous *theorem* by George Birkhoff, which we will discuss in Chapter 8). However, it is unlikely that Schwarzschild himself fully appreciated the disruptive novelty or the implications of his solution. Like many scientists after him, he must have considered it an 'oddity' of the new theory, to the point that he quickly entered into intense correspondence with Einstein, asking for his opinion.

There are several aspects of Schwarzschild's solution that make it seem rather bizarre to say the least. First, it is spherically symmetric, and therefore it has a 'centre' and a 'far-away' region (also referred to as the *asymptotic* region). In this region, and thus at large distances from the centre, the Schwarzschild solution is not particularly strange, being the same as that of a source of gravity in Newton's theory of gravity. However, while in Newton's theory, this result is possible only in the presence of matter (since an object with mass has to produce the gravitational field), the Schwarzschild solution is, as mentioned above, in a vacuum. In essence, it suggests that it is possible to create a gravitational field from . . . nothing! A decidedly extravagant idea.

A second oddity is that it predicts the existence of a spherical surface with radius $r = r_S$ whose value is given by the following equation:

$$r_S = 2M \frac{G}{c^2} \tag{6.1}$$

Exactly at this radius, also known as the *Schwarzschild radius* and representing one of the fundamental length-scales in astronomy, a number of strange phenomena occur, which we will discuss shortly. However, the most peculiar and disturbing of these phenomena is that the Schwarzschild solution is not well defined at this radius. From a mathematical point of view, the functions that describe the solution at that radius *diverge*, that is, they assume infinitely large values.

It is evident at first sight that Equation (6.1) tells us the Schwarzschild radius is equal to twice the mass M of the object in question: the black hole (again, I have greyed out the constants that do not have a 'physical meaning'). Today we know that the Schwarzschild radius r_S represents the radius of the so-called *event horizon* (more on this later) and that the divergence of the solution at this radius is not particularly worrying; it is simply the result of an inadequate choice of coordinates. Indeed, it is possible to use a whole set of better-suited coordinates that 'eliminate' this singularity and make the radius $r = 2GM/c^2$ perfectly regular, that is, without divergences.

However, other peculiarities remain at the Schwarzschild radius. In particular, it is impossible for a ray of light emitted at this radius (but also for any object with mass) to move outwards to larger radii, although the ray can move inwards. In other words, the spherical surface with a radius equal to the Schwarzschild radius behaves as a unidirectional membrane, allowing the absorption of objects, but not letting anything escape. Later on, we will see what this means from a physical point of view.

Finally, a third oddity of the Schwarzschild solution – one with the most serious complications – is that it also diverges at the centre, that is, at $r = 0$. However, unlike the singularity at the horizon, this one in the centre is much more severe and cannot be 'healed', even through an intelligent transformation of coordinates. Instead, all the possible indicators of what happens when one approaches the zero radius suggest that every physical and geometrical quantity becomes infinitely large there. For this reason, $r = 0$ has the fascinating but somewhat disturbing name of *spacetime singularity*.

Today, we refer to the point $r = 0$ in the Schwarzschild solution as a *physical singularity*, that is to say, a point of spacetime where the equations of general relativity are impossible to solve because they are divergent. In turn, this means that near the singularity, the laws of physics cease to be valid, and we are therefore unable to predict or even understand what might happen there. The physical singularity still remains the biggest thorn in the side of Einstein's theory: the clear manifestation of the fact that, however correct the theory may be, it remains incomplete. Indeed, although general relativity has been proven to provide correct predictions on the behaviour of gravity on large scales, it remains unable to describe gravity in the presence of very small dimensions and large curvatures, that is, scales at which the laws of quantum mechanics come into play. Even today, despite the efforts of generations of theoretical physicists, we lack a theory of *quantum gravity*. But, fear not, we will return to this point at the end of our journey.

So, when proposed in 1916, the Schwarzschild solution triggered mixed reactions. On the one hand, it showed that the Einstein equations were extremely interesting and had solutions and, on the other hand, it raised more questions than answers. As you can imagine, given its oddities, perhaps Schwarzschild's solution was a poor way to begin to understand the consequences of general relativity. This may be why the scientific community of those years received the new theory proposed by Einstein with an attitude of detached interest and sceptical curiosity. This was true for the mathematicians, who were attracted by the theory's geometrical implications, and the physicists, who were instead interested in the experimental impacts.

The scientific climate of those years was characterised by a courteous appreciation of the theory as well as a fundamental indifference to it, made worse by the difficulty in finding solutions and thus understanding the predictions. Hence, the theory of general relativity became the 'victim' of a widespread lack of interest on the part of the scientific community. However, such a feeling was not unreasonable and was in part fuelled by some important 'distractions'. At that time, an equally compelling theory had begun to attract the attention of the physics community – *quantum mechanics* – another fascinating theory seeking to unveil the properties of elementary particles. This new theory was also bizarre but had the major advantage of offering much easier predictions to understand and, above all, to verify through experiments.

As a result, during the first years of its existence, Einstein's theory seemed doomed to slip slowly into oblivion. Fortunately, the advent of X-ray astronomy and the observation of objects such as Cyg X-1 forced the scientific community, 50 years later, to reconsider its predictions.

Escape Velocity, Event Horizon and Singularities

Having covered the main oddities presented by the Schwarzschild solution, we will now try and understand them. In particular, we will discuss the event horizon, the peculiar properties of this mathematical surface and, finally, the physical singularity at the centre of a black hole.

First, to illustrate the features of the event horizon, it is helpful to start from a phenomenon that we can experience directly. Let us imagine taking a tennis ball and hitting it upwards, being careful to give it a trajectory that is as vertical as possible. We will observe that the ball will decelerate given a certain initial speed. More precisely, its speed will progressively decrease during the ascent until it becomes zero at a precise point of its vertical motion. At that point, the ball will stop rising and fall back, accelerating as it returns towards you. We know that the maximum height reached by the ball is directly linked to the force with which we have hit it. Indeed, if we relaunch the ball with a different speed, the maximum height reached when the motion is reversed will change. In particular, this height will be lower if the initial speed is lower, and higher if the starting speed is higher. Therefore, it is clear that there is a precise relationship between the initial speed with which we hit the ball and the maximum height it reaches. In physics, we like to think that the height represents the position where the ball's initial kinetic energy (the energy it has because of its motion) is converted entirely into gravitational potential energy, that is, the energy it has because of its position.

The most important aspect of this simple example is that there is a reason why the tennis ball falls back, and this is because it is *gravitationally bound* . In other words, the gravitational field produced by the Earth is so dominant over the dynamics of the ball that it is unable to escape from it. Following this line of thought, we may ask ourselves if we could use an initial speed that would allow the ball to escape the gravitational field or become *gravitationally unbound*. The answer is yes and is what we call, unsurprisingly, the *escape velocity*. Calculating its value is quite simple and requires comparing the kinetic energy with the gravitational potential energy. In this way, we obtain an expression that is valid in both Newton's and Einstein's theory of gravity, namely:

$$v_f = \sqrt{2\frac{M}{R}G} \tag{6.2}$$

The expression above tells us that, given a source of a gravitational field with mass M and radius R (for example, the Earth), the escape velocity is proportional to the square root of the mass and inversely proportional to the square root of the radius of the source. The escape velocity grows with the mass of the planet (the more massive the latter, the greater the speed necessary to escape it), and it decreases as the size of the planet increases.

Putting down some numbers, if we wanted to hit our tennis ball so that it escaped the Earth's gravitational field, we would need to launch it with a speed of 11.2 kilometres per second, or 40,320 kilometres per hour. A speed so high that not even the best tennis player could ever send a ball into orbit on our planet, which is quite fortunate as it would be very frustrating to watch a tennis match in which players cannot hit the ball too high for fear of losing it!

That said, Equation (6.2) is interesting for at least two reasons. The first is that the escape velocity is independent of the mass of the ball we are hitting (only the mass of the source of the gravitational field appears in the formula). In other words, to send either a grain of sand or a spaceship to the Moon requires the same velocity (obviously, the energy spent to do so would be very different . . .). The second reason is that, as you may have noticed, the escape velocity depends on a quantity that we have already encountered: the compactness M/R of the source of the gravitational field. This means that different celestial objects (the Earth, the Moon, the Sun . . .) will have different escape velocities because their compactness and, consequently, their gravitational fields are different. Indeed, the escape velocity on the Moon is almost five times smaller than on Earth – about 2.4 kilometres per second. Hence, even on our natural satellite, we would not be able to hit the ball out of the gravitational field with the simple force of our muscles, but we would undoubtedly hit it much higher than on Earth. Conversely, if we tried to do this on the Sun, it would be difficult to hit it upwards at all, since the escape velocity there is approximately 615 kilometres per second.

Incidentally, the very fact that the escape velocity on the Moon is so low – lower than the average velocity of the molecules of nitrogen gas – explains why the Moon does not have an atmosphere. Luckily for us, the Earth's escape velocity is large enough that it has retained an atmosphere, the presence of which is essential for life on our planet.

When the Escape Velocity Is as Large as Possible

There is another aspect of Equation (6.2) worth underlining. We know there is a maximum speed at which objects can move: the speed of light. So, we can use this equation to calculate the compactness M/R at which the escape speed velocity is equal to the speed of light. If we do that, we will discover that, ignoring the constants G and c^2, the value of the compactness is $M/R = 1/2$. This result makes the Schwarzschild radius, as given by Equation (6.1), much easier to understand: it is the radius of an object, more specifically a black hole, whose compactness is $M/R = 1/2$, which is the maximum compactness seen so far. This allows us to introduce a new definition of the event horizon as the surface at which the escape velocity is the maximum possible: that of light. In other words, if emitted exactly from the event horizon, light would not be able to leave the gravitational field of the black hole because it would need an escape velocity greater than the maximum allowed by physics.

Imagine now being on this mathematical surface and emitting a laser beam towards a distant star. Due to the very strong gravitational field, the beam cannot propagate outwards and instead will have to go back! I should emphasise that this behaviour would only happen when sitting exactly on the event horizon. If we placed ourselves a little further out and repeated the experiment, the gravitational field would still be enormously strong, but part of our laser beam would be able to escape it and reach an observer at a large distance. Put another way, this peculiar behaviour of light applies only as long we are at $r = r_S$; outside the event horizon, that is, at radii larger than r_S, the gravitational field remains extremely intense, but it is not 'invincible'. Conversely, if we emitted a beam of light within the event horizon, that is, at radii smaller than r_S, the beam would only move inwards, thus travelling towards the centre of the black hole where the physical singularity is located.

Nevertheless, let us go back to considering the beam of light emitted just outside the event horizon. We have seen that part of this beam would reach an observer even at a considerable distance, for example, on Earth. However, the light perceived by such an observer would be very different, in terms of colour (or wavelength), from that emitted initially. Let's assume that, when emitted, the light was at a visible wavelength, say in the *blue* part of the spectrum. The observer at a large distance would instead receive it in the *red* part of the visible spectrum!

Furthermore, this tendency of the light to 'appear red' would be stronger the closer the emission is to the event horizon. The wavelength of a light beam tells us about the energy of the photons it is composed of; to be more precise, the energy of a photon is inversely proportional to its wavelength. This process of the reddening of the light reveals that the light beam loses energy when propagating from its emission point, near the event horizon, to the distant observer.

This seemingly bizarre process actually makes perfect sense if we go back to thinking about our tennis ball being hit upwards. Also, when moving upwards, the ball loses kinetic energy, which transforms into gravitational potential energy. At a certain point, the ball is forced to stop exhausting all its kinetic energy and has to fall back. Since a photon cannot slow down because, by definition, it is forced to move at the speed of light, it manifests this loss of energy by changing its wavelength. I have just described a well-known phenomenon in astrophysics, which reflects a very common property of the emission from compact celestial objects, namely, the *gravitational red-shift*. This phenomenon is most extreme precisely in the case of black holes and becomes mathematically infinite for a photon emitted from the event horizon. If the red-shift of a photon emitted at $r = r_S$ is infinite, the observer will never receive it. Hence, the infinite red-shift experienced by a photon emitted at the event horizon is a different way of understanding why a photon emitted at the event horizon cannot propagate outwards but can, at most, remain on orbits tangent to this surface.

To make this concept as intuitive as possible, let's perform a thought experiment together. Imagine yourself at some distance from the event horizon – perhaps aboard a spaceship positioned on one of the stable circular orbits (that are allowed under the laws of general relativity) around a black hole – and dropping a device similar to a light on a buoy at sea, flashing at regular intervals, towards the celestial body. Well, as the buoy moves away from you and approaches the event horizon, you will see that the interval between each flash increases, while the light produced by the device, initially an intense white, will turn more and more to red. This process will continue as long as a sufficient number of photons can reach you but, as time goes by, you will receive fewer and fewer photons, and these will shift more and more to red. At some point, when the buoy is very close to the event horizon, you will no longer see any light, no matter how sensitive your instruments are. However, this does not mean that the device stopped working or was destroyed before crossing the event horizon! If you had been attached to the buoy, you would have seen that its flashes always had the same periodicity, and the light did not change colour at all. Furthermore, after a precise time interval that is simple enough to calculate, you would cross the event horizon and find yourself inside it, unable to communicate with the universe outside. Depending on the mass of the black hole, and especially if it were supermassive, the passage through the event horizon would not even be accompanied by any particularly dramatic event.[2]

This difference in the *perception* of physical processes by different observers (i.e., you observing the buoy from the spaceship; and you falling together with the buoy) is a genuine property of general relativity – from which its name derives. Although this 'relativity' of perception may seem unsettling, it disappears when one does not discuss observational perceptions but the physical processes *themselves*. When one does that, all observers will agree on the same conclusion.

The example of the buoy teaches us that it is impossible to observe an event horizon (or the birth of a black hole) *directly* as long as such observations involve photons and, thus, the emission and reception of light. However, it is still possible to obtain *indirect* evidence of the presence of a black hole, either through the emission of electromagnetic radiation from its vicinity (as we will see in Chapter 7) or even without any light and through the emission of gravitational waves (as we will learn in Chapter 8).

At this point, I am aware that I have provided you with a lot of information about black holes and their event horizons. So, let me briefly recap to put some order and focus on the more essential points we have covered so far:

- It is possible to shine a beam of light outwards from a point very far outside the event horizon: it will propagate, reaching regions placed at larger distances without perceptibly changing its wavelength.
- It is possible to shine a beam of light moving outwards from a point close to the event horizon but still outside it, but part of it will bend backwards and fall onto the horizon. As the emission of light takes place closer and closer to the event horizon, the amount of light that fails to propagate outward becomes larger.
- The light emitted near the event horizon will change wavelength, turning red due to the gravitational red-shift effect. The strength of this red-shift increases as its emission is made from points closer to the horizon and becomes infinite for a photon emitted exactly at that surface.
- A beam of light emitted from the event horizon, that is, a radius $r = 2M$, cannot propagate outwards: if the black hole does not absorb it, its destiny is to remain forever tangent to that spherical surface.
- Unlike that of a neutron star, the event horizon is not a solid surface but a purely *mathematical* one. It is the boundary marking the transition between two physical processes: one where a photon manages to propagate outwards and another where the photon is instead forced to 'return'.
- Much of what is said above for photons is also true for objects with a mass, which, on approaching the event horizon, will need larger and larger energies to escape the

gravitational field. The energy required becomes infinite if the particle is placed exactly on the event horizon.

I will conclude this discussion on a historical note. I have already mentioned that the expression for the escape velocity (Equation (6.2)) is also the same within Newton's theory of gravity and was already known from the beginning of the eighteenth century. Indeed, the British naturalist John Michell (1724–1793) was the first to formulate what could be considered an early idea of an event horizon. In an article published in 1784,[3] he introduced the concept of 'dark stars'. In his line of arguments, these would be celestial bodies whose compactness M/R would be such that the escape velocity was equal to that of light. Hence, such objects would not be able to emit any light and would therefore be dark stars. Of course, in 1784, general relativity was still to come, as was the concept of the event horizon, and observations had not yet revealed the presence of objects such as Cyg X-1. Yet, it is quite incredible that ideas linked to black holes were formulated (albeit in a somewhat embryonic form) centuries before these concepts were put into sharp focus.

When Time Makes Fun of Us

It is now time to face another physical phenomenon near the event horizon, which is recurrent in all science-fiction films involving black holes: the *dilation of time*.

In Chapter 4, we saw how perfectly plausible it is that if mass or energy can bend space, the same should also happen with time. Furthermore, we have said that the wavelength of a photon emitted near an event horizon and directed outwards is stretched out by gravity, causing what is known as gravitational red-shift. So, now we just have to put these two concepts together.

To do this, I will use two hypothetical observers, Anna and Emilia, each onboard their own spaceship orbiting a black hole on a stable circular orbit so that neither of them runs any risk of falling into the black hole. Rather, they will maintain a constant distance from the black hole, with Anna closer to the event horizon and Emilia at a larger distance from it. Now, suppose that Anna emits a beam of blue light directed towards Emilia: as we have seen, the latter will receive it without any problem but at a reddish wavelength. We have interpreted this phenomenon as the consequence of energy expended by the photon on exiting the gravitational well from which it was emitted. However, there is another way to describe the same phenomenon. A photon is an electromagnetic wave, and the number of crests associated with such a wave must be the same for Anna, who sent it, and for Emilia, who received it. Furthermore, since the photon must move at the speed of light, the change in wavelength can be interpreted as a variation of the progress of time measured by Anna and Emilia. In short, the fact that the two observers register different frequencies for the same photon (remembering the frequency f of a photon is inversely proportional to its wavelength $\lambda : f = c/\lambda$) can be interpreted as the consequence of the fact that the time taken to count the same number of crests is different for each observer.

We have encountered something similar in Chapter 4 when discussing the slight difference between clocks placed on the ground and clocks placed on satellites. Well, as with that phenomenon, the gravitational red-shift experienced by Anna and Emilia can be described as the manifestation of the elasticity of time. When comparing what happens with the GPS, the important difference is that because Anna and Emilia are close to a black hole where the curvature is extreme, they will register a much larger (possibly extreme) variation in the progress of time. Indeed, this difference will diverge near the event horizon. Put another way, the almost infinite red-shift suffered by a photon emitted near the event horizon can be interpreted as being due to the enormous 'slowdown' of time as the photon tries to move outwards. To clarify, the observer at a very large distance from the black hole will find time progresses at the same rate it has in a flat spacetime. In contrast, for an observer near the black hole, time progresses more slowly since that region of spacetime is far more curved. Ultimately, time would virtually freeze for an observer on the event horizon.

Perhaps we need to make use of another thought experiment to understand this concept better. However, we will have to stretch the imagination a little further this time as we find ourselves using extreme, albeit perfectly plausible, physical conditions.

So, let's imagine two astronauts, Carolin and Dominik, onboard a spaceship in a circular orbit at a safe distance from a supermassive black hole, for example, the one at the centre of the elliptical galaxy M87 (I will elaborate on this fantastic object in Chapter 7). The two astronauts are bored and decide to take a 'space trip' on the smaller service spacecraft. Since much of the action in that area occurs around the black hole (one way or another, it seems that everything literally ends up falling into it!), they decide to get closer to it while remaining at a safe distance from the event horizon. Both are well aware that there are safe orbits for the small spacecraft, which can be calculated based on the power of their engines. After checking everything is in order, they write a note in the electronic logbook, mark the exact coordinates, including the astral date and time, and leave the mothership.

The gravitational fields in the vicinity of a supermassive black hole, such as M87* (which is found at the centre of the galaxy M87), vary very gradually. To be more precise, the curvature radius near these objects is extremely large, and differences between nearby locations (i.e., the tidal fields) are quite small. So, instead of turning on the spacecraft's rockets, Carolin and Dominik decide to simply enjoy their 'free fall' as they are pulled towards the centre of M87. It is as if Carolin and Dominik were gradually descending a mountain with enormous horizontal dimensions and a very gentle slope. However, in reality, the gravitational attraction of the black hole is so great that, even without their engines, Carolin and Dominik soon reach very high speeds, close to that of light.

Meanwhile, sipping hot tea and tucking into fresh pastries, the two astronauts delight in looking through the spacecraft windows at the breathtaking panorama of the matter as it converges with them towards the black hole. Then finally, they decide it's time to go back and the instruments confirm that, at their distance from the event horizon, they can still reverse their course. So, they turn on the engines and point the spacecraft towards the mothership, employing, it has to be said, a considerable amount of energy to ascend the gravitational well into which they had gradually descended. In essence, the climb away from the black hole is the difficult part . . . .

Once they reach the mothership, they check the instruments. The calendar marks a very different astral date from that on which they left and that of their small spacecraft – according to the computer on the mothership, their trip lasted several months. Carolin and Dominik, however, are not surprised; they knew this would happen, which is why they took the fresh pastries with them rather than leaving them behind in the ship's fridge. Of course, time passed much faster there than it did onboard the service spacecraft, but it was a perfectly normal and totally predictable phenomenon for the two astronauts. Actually, before leaving, Dominik had calculated the time they would 'measure' onboard the small spacecraft to reach the event horizon or, had they decided to cross it, the physical singularity at its centre. Aware of the difference between 'their' time on the small spacecraft and that measured by the instruments onboard the mothership, he calculated the route, ensuring that the difference between the two was not excessive: he did not want the mothership to remain without a crew for too long. In short, this trip reveals that Carolin and Dominik are familiar with general relativity and the subtleties of time in strong gravitational fields.

That time can slow down and even stop seems absurd to us, so it can be helpful to reflect on two important aspects of this phenomenon. First, we must never forget that what seems implausible and even counter-intuitive to us can be very plausible if allowed by physics. Our intuition is built on experience accumulated on this planet, under specific physical conditions, that is, imperceptible curvatures, low temperatures and densities, etc. Therefore, to establish what is physically possible, we must rely on the equations that describe the laws of physics and do so almost blindly. If a theory is correct and its equations allow a certain process to take place, then that process is plausible (although not necessarily *probable*).

The second aspect to keep in mind is that time dilation is not a science-fictional invention but a reality on Earth supported by considerable experimental evidence. In fact, we test its correctness every time we use a satellite navigator. When compared to Carolin and Dominik's experience on their spacecraft, the only difference is the magnitude of this phenomenon, which is enormously amplified in the vicinity of a black hole or a neutron star. After all, it is precisely for this reason that such objects fascinate us so much!

How Do You Produce a Black Hole?

Now that we have a better idea of what a black hole is and what its most salient features are, it is time to ask ourselves a basic but fundamental question: 'How are these objects produced?'.

The answer is not exactly simple but not too complicated either, especially if you leave out a few details, which is just what I intend to do in this section. Furthermore, I will limit the discussion to black holes with a mass of between about 3 and 100 solar masses (typically called *stellar black holes*) as their formation process is much clearer than that of supermassive black holes, whose cosmological origin is currently the subject of debate, and because what I will say holds true in all cases, at least qualitatively. Also, I will take advantage of what we learned in Chapter 5 about neutron stars and employ an invaluable theorem to illustrate the process of black-hole formation.

But let's proceed in an orderly way. To explain how a stellar black hole is created, I need to recall an important theorem by the German mathematician (also a naturalised Australian citizen) Hans Buchdahl (1919–2010). The theorem is very simple, and its thesis is expressed as follows: given a self-gravitating object made of matter and having a mass $M_\star$ and a radius $R_\star$, its compactness $M_\star/R_\star$ is bound to be:

$$\frac{G}{c^2}\frac{M_\star}{R_\star} \leq \frac{4}{9} \qquad (6.3)$$

In other words, given a self-gravitating object, its radius $R_\star$ must be greater than $9/4 = 2.25$ times its mass $M_\star$. Note that the theorem does not place a limit on the mass (which can therefore be tiny or huge) nor on the radius needed to create a black hole. What it tells us, instead, is that for every mass, whatever it is, there is a *limiting radius* below which it is not possible to find an equilibrium solution to the Einstein equations. Furthermore, Equation (6.3) also tells us that this limiting radius is equal to 9/8 of the Schwarzschild radius, which, I recall, is $r_S = 2GM/c^2$. Therefore, there is another way to describe the implications of Buchdahl's theorem. It is impossible to arbitrarily reduce the radius of a compact object with a solid surface and still obtain an equilibrium solution. Instead, a limit is given by a radius that is only 13% larger than that of the event horizon of a black hole with the same mass.

The proof of the thesis of Buchdahl's theorem proceeds 'ad absurdum', namely, by showing that if an object made of ordinary matter, that is, with a positive energy density and pressure, were built with a compactness greater than 4/9, then the pressure at its centre would be infinite. And since the pressure must necessarily be finite, it follows that such an object cannot exist. Later on in this chapter, we will see that, in principle, it is possible to build objects with an even greater compactness. First, however, we need to renounce an important assumption in the theorem and rely on a decidedly *exotic* type of matter.

This beautiful theorem offers us the opportunity for another of our thought experiments. We can imagine taking an object with a mass $M_\star$ and progressively reducing its size in order to increase its compactness. As I am sure you recall, we did something like this in Chapter 4, when we imagined compressing the Sun to bring its radius to only five kilometres and then calculated the resulting curvature. The experiment I am proposing here is similar, but only in part because now we want to learn what happens when we exceed the limit imposed by Buchdahl's theorem.

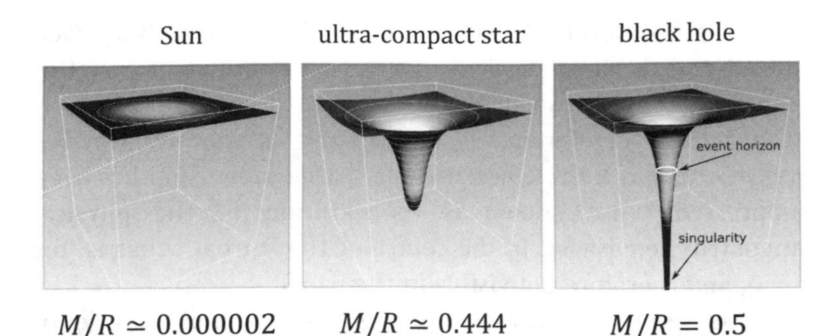

Fig. 6.1 The curvature of spacetime produced by the Sun, on the left, is created when we reduce its radius while maintaining its mass. The square in the centre depicts the Buchdahl limit while, in the one on the right, its radius has been brought to just 3 kilometres, generating a black hole. A black and white version of this figure will appear in some formats. For the colour version, please refer to the plate section.

To help us in this thought experiment, we can use the three boxes in Figure 6.1, which show the curvature of spacetime produced by the Sun as we reduce its radius while keeping the mass stable. Below each image, I have indicated the relative compactness *M*/*R*. The box on the left shows the curvature produced by the Sun with its true radius of about 700,000 kilometres. As seen in Chapter 4, the compactness is of some parts in a million, and the curvature is almost zero, so the surface appears flat. On the other hand, the middle panel shows the curvature produced by compressing all the atoms of the Sun into an ultra-compact star, that is, to the limit set by Buchdahl's theorem. In this case, the compactness is exactly $M/R = 4/9 \simeq 0.444$, and the curvature is clearly visible: a depression with very steep walls has developed in the centre of the surface.

Finally, the panel on the right shows what would happen if we took the ultra-compact star and compressed it further, even if only slightly, always keeping the mass unchanged. This would produce an object with a spherical surface having a radius of only 3 kilometres and a compactness of 0.5. That is, a black hole! When we defined the Schwarzschild radius as being equal to twice the mass of the black hole in Equation (6.1), we also implicitly stated that the compactness of a Schwarzschild black hole is exactly equal to $M/R = 1/2 = 0.5$. In the right panel of Figure 6.1, I have shown the position of the event horizon as given by a circumference of radius $r = r_S$ (a spherical surface appears here as a circle because one spatial dimension has been suppressed). Also shown is the position of the physical singularity, envisaged in the centre of the 'throat', that is, for $r = 0$, and therefore not visible in the panel.

As you are now accustomed to doing, we can make a series of considerations inspired by the sequence in Figure 6.1. The first consideration is that we can use the panel on the right (the black hole) to introduce a mechanical analogy to help you understand the role gravity and the event horizon play. Imagine that the curved surface represents the bed of a shallow lake, with a hole in its centre. The presence of the hole will create a current that will drag the water of the lake towards the bottom, creating a huge waterfall (let's ignore for the moment the presence of rotation, so the water does not produce a vortex but falls along radial trajectories, just like a circular waterfall). At this point, picture yourself on a boat with extremely powerful engines. As you get closer to the waterfall, the current increases, but the engines on your boat are powerful enough to push against the intense current, allowing you to approach not only the outer edge of the hole but to go beyond the edge, even if the walls are almost vertical. Well, this is possible only up to a certain point. In fact, there will be a 'border' beyond which your engines will not be able to push you upstream, no matter how powerful they are. Indeed, even if their power were almost infinite, it would not be enough. Hence, for you and your boat, such a limit represents in all respects a 'horizon': a surface you can cross but not leave!

The second consideration is that there are no intermediate solutions between the middle and right panels in Figure 6.1. In other words, if we compress, even infinitesimally, an ultra-compact star that has reached the Buchdahl limit, the Einstein equations do not allow for any static solution other than that of the black hole. Therefore, a sharp transition is created between the solutions of objects composed of matter and having a solid surface, that is, compact stars, and those without matter and without a solid surface, namely, black holes.

A third consideration, possible when looking at the curvature shown in the last panel, is that it seems as if spacetime has been 'torn'; as if the very fabric of spacetime has given way under the irresistible action of gravity. But, in reality, spacetime cannot be torn apart since it is a regular continuum everywhere – with the sole exception of the physical singularity – and we should instead think that a black hole pushes its elasticity to the maximum. In other words, although we have seen that spacetime is generally quite rigid, in the case of a black hole, gravity is so strong that its deformability reaches its maximum value.

A fourth consideration emerges from the fact that in going from an ultra-compact star to a black hole, we have essentially concentrated all the matter in a single infinitesimal point. Therefore, a black hole is 'empty' since all the mass used to generate it is now concentrated in a point of zero dimensions. It follows that, with the exception of a point at the centre, the interior of the event horizon is empty! Therefore, those statements suggesting the existence of stable matter inside a black hole – unfortunately discussed all too often – are incorrect. The density of matter in a black hole is precisely *zero* because a black hole is a solution of the Einstein equations in a vacuum.

To conclude, let me dedicate a final consideration to the timescale over which a black hole is produced. I think it is now clear that nothing in physics happens instantaneously, which also applies to the birth of a black hole. Equally clear is that, within general relativity, this timescale inevitably depends on the position and motion of the observer. Someone far from an event horizon will never see an object fall on it. Likewise, someone far from a collapsing star will never see its surface disappear below the horizon itself. Rather, the observer will see a star progressively contracting while emitting light that gets dimmer and dimmer and shifts more and more towards the red until, when the emitted radiation is too weak to be received, it disappears. Instead, if the observer were to fall together with the collapsing matter, the precise duration of the collapse would depend in part on a series of factors, such as the type of matter or its rotation. However, omitting unnecessary complications from the start of the collapse and, therefore, from when gravity dominates every other physical interaction, the collapse essentially takes place over the free-fall timescale. Calculating this timescale is very easy, employing a simple formula that can be used for any object, large or small. Basically, the formula says that the free-fall timescale is inversely proportional to the square root of the density of the collapsing object. The denser the collapsing object, the shorter the free-fall time. In the case of the Sun, our favourite example, the free-fall time is about 27 minutes. To put it another way, if we could magically 'turn off' all thermonuclear reactions inside our star and allow it to collapse by removing all of its internal pressure, then the solar surface would reach the centre in a little less than half an hour. A surprisingly short time if we consider that matter on the surface has to cover the 700,000 kilometres separating it from the centre of the solar core.

So far, all of the considerations are meant to help us imagine what must have happened about five million years ago when the supernova explosion of a very massive star (over 40 solar masses) gave birth to the black hole in Cyg X-1. Indeed, the most relevant difference between what happens with a supernova that gives life to a neutron star and one that generates a black hole is that, in the latter case, the iron core is much more massive and of the order of about ten solar masses. Under such conditions, the density reached during the collapse is such that no pressure (not even that of the degenerate matter discussed in Chapter 5) can counteract the eventual collapse of the stellar core. Thus, when a sufficiently large amount of mass is concentrated within a radius of the order of the Schwarzschild radius, a black hole will inevitably have to be produced.

Nowadays, thanks to the numerical solution of the Einstein equations and those of relativistic hydrodynamics, we know quite precisely what happens in the process of the formation of a black hole. Around 2005, together with European and

American colleagues, I myself studied this process, producing the first numerical simulations to describe it in three dimensions, albeit when starting from some idealised initial conditions. Despite these simulations, many aspects of the astrophysical genesis of black holes, both in the case of stellar and of supermassive black holes, remain unclear. In the case of stellar black holes, electromagnetic observations tell us they exist with masses between 5 and 10 solar masses. In contrast, the detection of gravitational waves so far (which we will discuss in Chapter 8) suggests that they exist with masses between 10 and 80 solar masses. The situation becomes even more uncertain when we consider supermassive black holes, that is, black holes of millions or more solar masses. There is far greater uncertainty about the mass and cosmological era in which these are first produced. However, it seems clear that, even when considering these enormous black holes, their birth must at some point in time involve the condensation of enormous quantities of matter into a region with a size comparable to the Schwarzschild radius. For this reason, the thought experiment conducted previously remains perfectly valid for all types of black holes and is a helpful reference, regardless of the mass of the black hole under consideration.

Black Holes Have No Hair

So far, I have always used mass as the only property for distinguishing black holes of different types and have discussed small (or stellar) and large (or supermassive) black holes. However, it is reasonable to wonder if such a complex object, described by equations that are extremely difficult to handle and even more so to solve, can really be described only in terms of its mass. In other words, it is time to ask ourselves: 'How complex are black holes?'.

To provide an answer, I need to digress briefly. As small children, our perception of reality (and consequently our understanding of the world) teaches us that a wealth of details inevitably accompanies complexity. It seems obvious that a natural phenomenon with a complex manifestation must be full of 'properties'. To be clearer, let me turn to a few examples. Let's start from a context known to all, at least on a general level: music. I think it is evident to anyone that a symphony of classical music performed by a large orchestra is much richer in content and, therefore, in nuance than the simple sound produced by a tuning fork. Because I am aware that speaking of 'properties' or 'richness of content' may appear qualitative and subjective, we should stick to a more objective measure. Therefore, consider the number of instruments involved in a symphony of medium duration and multiply it by the number of musical notes played during the symphony. This will give us a far more quantitative and objective measure of the complexity of the symphony. The same calculation can then be made for a tuning fork that repeats its note at regular intervals over the same period of time. Using these more objective terms, it is then trivial to conclude that the symphony orchestra is hundreds of times more complex than a tuning fork.

It is possible to find similar examples within a biological context. For instance, we could consider three living organisms we are familiar with: a bacterium such as *Escherichia coli*, a *Drosophila* fly, and an earthworm. These organisms are obviously characterised by an increasing complexity since more and more information is needed to describe them in this order. To quantify this complexity in numerical terms, we could consider their genetic makeup: 5,000 genes for the *Escherichia coli*, 13,000 for the *Drosophila* fly and 19,000 for the earthworm (the number of genes in a human is of the order of 30,000). Because each gene can be described in terms of a specific combination of amino acids, it is clear that an enormous amount of information is needed to describe even the simplest of living beings.

In essence, therefore, both our daily experience and our knowledge of the world suggest that the more complex an object, organism or physical phenomenon is, the greater the amount of information needed to describe it. This would all make perfect sense if not for the fact that black holes are a striking exception to this very reasonable rule.

Based on these examples, we could conclude that describing a black hole (the most extreme result of the most complex equations of theoretical physics) would require an enormous amount of information and that mass is just one of the innumerable properties of black holes. Well, quite surprisingly, that's not the case: in reality, black holes are the simplest macroscopic physical objects ever! This conclusion is the simple consequence that to describe any black hole we need only *three numbers*, which embody as many properties. Once again, the concept of a black hole surprises us – it is immensely easier to describe than even the most elementary bacterium.

So, what are the three, and only three, properties of a black hole? They are the mass, which we have already discussed; the *spin*, namely the amount of intrinsic rotation; and, finally, the electric charge. Generally, these properties are represented by M for the mass, J for the spin angular momentum and Q for the electric charge. Thus, we can assign any black hole a place in the 'database' of the universe by simply allocating to it the three values of M, J and Q. In a realistic astrophysical scenario, the electric charge of a black hole is expected to be essentially zero, so in practice, we can even forget about the charge. The reason for this is easy to understand. If a black hole with a net electrical charge of a certain sign was ever generated, the ever-present free charges, that is, the set of electrons, protons and other charged particles that circulate freely in interstellar space (some of which would necessarily have the opposite sign to the black hole's charge), would be electrically attracted by the black hole and absorbed. In this way, the charge of the black hole would quickly diminish until it became zero. At that point, the black hole would be electrically neutral (i.e., $Q = 0$), and therefore unable to attract any other charge. This explains why, in practice, astrophysicists use only two values, M and J, to describe these amazing celestial bodies.

The spin angular momentum must be within a well-defined range to simplify things. For this purpose, it is helpful to consider *dimensionless* quantities, that is, quantities that are defined in such a way that they have no physical dimensions (kilometres, kilograms, etc.) and are therefore pure numbers. In this case, we can define the *dimensionless spin* of the black hole, given by the ratio between the spin angular momentum and the square of the mass, namely, $a = J/M^2$. Well, for a black hole in general relativity this number must be between -1 and $+1$ (the difference in sign reflects the orientation of the rotation with respect to a reference axis and, thus, whether the black hole is rotating clockwise or anti-clockwise). The quantity a is zero in the case of a Schwarzschild black hole, which is, therefore, a *non-rotating* black hole (from here on, I will refer to it as either a *Schwarzschild* or *non-rotating* black hole). However, if the dimensionless spin a has a non-zero value, I will refer to a *rotating* black hole, or *Kerr* black hole (after the New Zealand mathematician Roy Kerr, who first found the solution[4]). As black holes typically accrete matter that has a certain amount of angular momentum, in an astrophysical context, black holes are always expected to be of the Kerr type, even though some may be only slowly rotating.

The peculiar conclusion that only three properties describe a black hole is contained in the famous so-called *no-hair theorem*. The playful name, given by the aforementioned John Wheeler and the Italian astrophysicist Remo Ruffini,[5] is aimed at underlining that if a black hole is seen as a sphere, then it is devoid of details: just like one's head can be devoid of hair. The thesis of the theorem, which is very simple to formulate, essentially states that in general relativity a black hole that is isolated and in an asymptotically flat spacetime (i.e., becomes flat at great distances) can be described by just these three values: M, J and Q. Interestingly, should something intervene to modify one of these numbers, for instance, due to matter with angular momentum falling onto the black hole and changing the values of M and J, the variation of the parameters will be followed by a state in which all transient perturbations are removed, and a 'new' black hole remains, although different from the previous one, still described by values M, J and Q (we will return to this point in Chapter 8).

Cosmic Engines and Hawking Radiation

In many respects, black holes can be seen as '*cosmic engines*' capable of absorbing matter from the outside and thus modifying their mass, angular momentum and electric charge (see Figure 6.2). Indeed, three laws of 'black-hole dynamics' exist that tell us how, through interaction with the external world, the properties M, J and Q can change over time. Leaving aside the charge Q, whose evolution we have already discussed, these laws (which derive from a classical context of general relativity without allowing for quantum-mechanical effects) tell us that the mass of a black hole can only increase. In contrast, its angular momentum can increase or decrease, possibly becoming zero. In essence, a rotating black hole can slow down and thus become a

Fig. 6.2 No-hair theorem: regardless of the type of objects that black holes can enhance, they only produce a change of the three numbers that characterise a black hole: the mass M, the angular momentum J and the electric charge Q.

non-rotating hole through the loss of its rotational energy. When this process occurs, a part of the mass of the black hole may actually be reduced, but the total mass cannot be reduced entirely. More precisely, the part of the mass associated with rotation (the part that depends on J) can indeed decrease; for obvious reasons, this part of the mass is also called the *reducible mass*. However, there is a part of the mass of the black hole not tied to rotation (i.e., which does not depend on J) that cannot decrease, but only increase. Again, for obvious reasons, this part of the mass takes the name of *irreducible mass*.

The law according to which a black hole, in the classical formulation of the theory of general relativity, cannot lose its irreducible mass was revolutionised in the 1970s by the British physicist Stephen Hawking. He proposed a new mechanism related to the principles of quantum mechanics that indeed violates this law.

Hawking suggested that a black hole could 'evaporate' and, therefore, lose mass (reducible and irreducible) through the emission of a thermal type of radiation, called *Hawking radiation*.[6] To produce such radiation requires the generation of a series of virtual particles and antiparticles (i.e., particles with 'negative mass/energy') in an extremely thin region near the event horizon. We know that particles and antiparticles can be produced anywhere in spacetime, with no net effect since the number of virtual antiparticles produced is balanced precisely by that of the corresponding particles. However, when the phenomenon occurs near an event horizon, which can only absorb particles but not emit them, a net result is possible and the overall effect is that the black hole emits radiation in terms of these virtual particles. At the same time, the black hole's mass decreases due to the accretion of antiparticles, which have negative mass–energy and therefore contribute to a negative mass.

There is no experimental proof of the existence of Hawking radiation. Of course, the main reason for this is that it is impossible to create a black hole in the laboratory (although it is possible, in principle, to generate objects with similar properties, called black-hole 'analogues'). But this is not the only reason:

Hawking radiation is, in fact, very weak, both in the case of stellar black holes and even more so for supermassive black holes. More precisely, *Hawking's luminosity*, i.e., the amount of mass–energy lost by a black hole per unit of time due to Hawking radiation, is inversely proportional to the square of the black hole's mass. It follows that even a 'light' black hole, such as a black hole of one solar mass, would take about 10^{67} years to evaporate completely. Considering that the universe is about 10^{10} years old, it is clear that a black hole of this mass would take billions and billions and billions and billions of times the age of the universe to evaporate. The timescales become even larger if we consider more massive black holes, such as those at the centre of our galaxy. That is why the best opportunity to observe Hawking radiation comes from extremely small black holes. In particular, a black hole that could have been generated in the early universe with a mass of 10^{11} kilograms (that is, one billionth of that of an asteroid) would emit enough Hawking radiation to be on the verge of evaporating.

Despite this lack of observational evidence, the scientific community is unanimous in believing that Hawking radiation is physically possible and that, at least in principle, a black hole can also lose its mass, possibly disappearing completely or leaving behind a very small singularity.

In summary, when talking about mass, black holes seem to behave just like humans: they can sometimes 'lose weight' if they slow down but, in the vast majority of cases, and sooner or later, they inevitably tend to put on weight....

A Difficult-To-Digest Idea

The 'absence of hair' in black holes presents a starting point for important reflection on how the very notion of these objects, and particularly the existence of an event horizon, is still hard to accept. An event horizon makes a black hole decidedly 'exceptional', in the sense that it represents an *exception* to what we know and are accustomed to accepting in physics. Once the mathematical surface of the event horizon is crossed, in fact, all information about the properties of the object entering the black hole is simply lost; converted, together with all of its complexity, into the only three properties an external observer can measure: M, J and Q.

If you think about it, this behaviour is anomalous, to say the least, and certainly not what we are used to in physics. Let me give you an example that illustrates how exceptional this behaviour is. Imagine taking a few logs of wood and lighting a nice bright fire; when the flames die down, all that remains of the wood are the fine indistinguishable ashes. You might not think these ashes contain any information about the original properties of the wood or the chemical properties of the organic cells that constituted the wood. In essence, you would think it impossible to recover any traces that would enable you to establish whether the wood logs came from a beech tree or an oak tree. Any information would seem to have gone up in smoke . . .! Well, that's precisely the point! The ashes are not the only product of the fire, and it is possible, at least in principle, to trace the properties of the wood by studying the gases contained in the smoke produced during the combustion.

This example shows that while systems in physics can undergo transformations, the complete information about the properties of the system is never lost. Of course, these properties can (and do) change, and it is quite possible that it can become increasingly difficult to collect information on such properties (think of the smoke, for example). However, for reasons at the very core of our understanding of physics, the information on the system is never lost. The only exception to this basic logic is what happens concerning a black hole! In this case alone, we lose all information about the properties of what generated the mass, the angular momentum and the charge of the black hole, even though we had all this information right up until the event horizon was crossed.

The inability to extract information from a black hole is called the *information-loss* paradox and is much more serious than it might appear. In fact, it undermines the very foundations of quantum mechanics, where the set of information in a physical system cannot be lost. More precisely, this paradox arises because, when describing the state of a system through a wave function (remembering a fundamental postulate of quantum mechanics is exactly that the complete information of a physical system is contained in its wave function), the evolution of this function must be unitary. In other words, having specified the initial conditions of the system's wave function, its evolution is fully determined. However, in the case of a black hole, this unitarity is violated so that several initially distinct states would merge into one and become indistinguishable.

The information-loss paradox remains unsolved, despite decades of studies on the subject. In fact, none of the theories introduced to remove it is convincing enough to be accepted – yet another strident confirmation of the incompatibility between classical general relativity, as formulated by Einstein, and the principles of quantum mechanics.

Giving Up a Cherished Method

Another aspect makes the event horizon a rather 'indigestible' concept for physicists. If you remember, in Chapter 2, I stressed the importance of Galileo's work in defining and introducing the scientific method – the only beacon illuminating human speculations. Over the centuries, Galileo's work has allowed us to reach firm conclusions in the debates, some lasting for centuries, that inevitably characterise scientific exploration: what is true and what is false, what is right and what is wrong. Galileo himself summarised the importance of this method in one of his best-known aphorisms. Turning to Tommaso Campanella, he declared: 'I prefer to find a truth, albeit in a small thing rather than to argue about the greatest issues without obtaining any truth'. In the case of a black hole, it is impossible to find a single truth, at least in terms of the scientific method, about what is beyond the event horizon, and this is not because the laws of physics cease to be valid!

Let me try and explain better what I mean. Imagine a pair of observers outside a supermassive black hole; Simplicius and

Salviati.[7] The two are arguing about whether or not it is possible to conduct an experiment within the event horizon. Following a logic of his own, Simplicius, who is not an expert in general relativity, affirms that, after crossing the horizon of events, there lies a fantastic world where no experiments are possible. Instead, everyone can get what they want there simply by thinking about it. According to Simplicius, if he crossed that limit, he would find a table laden with delicious foods and, in particular, his favourite ice cream (Simplicius loves to eat, and everyone knows he cannot resist ice cream).

For Salviati, who knows general relativity well, none of this happens inside the event horizon of a supermassive black hole. The laws of physics also remain valid there, and any object that has entered has to end up in the physical singularity, about which, he admits, he cannot make any prediction. Salviati also argues that if the black hole's mass is very large, then an object the size of a human being will experience very small tidal forces so that no appreciable change will be experienced when crossing the event horizon. Except, of course, for the fact that it will no longer be possible to communicate with the outside world.

Now, based on our understanding of physics, we can all agree that Simplicius' thesis is far-fetched. It has no logical or scientific support, and there is no reason to believe it to be correct. However, this is precisely the problem; the discussion between the two cannot be resolved, even if it is 'a small thing' in Galileo's terms. The simple reason is that neither Salviati, nor anyone else, can go inside the event horizon, verify that there are no tables laid with delicious foods and report the results to the rest of the world. Even more irritating for Salviati is that he knows very well it would be possible to *carry out* an experiment disproving Simplicius' thesis, but the problem would be communicating the results to the outside world since this is something you are unable to do once inside the event horizon.

Accepting the existence of an event horizon is therefore equivalent to admitting the presence, in our universe, of regions that are physically accessible but precluded from an investigation based on a scientific method. This is not because physics ceases to be valid there, since (at least in general relativity) the only problematic region is a point of infinitesimal dimension at the centre of the black hole. Therefore, much more trivial is the problem that the results of experiments conducted within the event horizon cannot be disclosed – a perspective that would certainly make poor Galileo turn in his grave . . . .

If we want, we can resort to a somewhat risky analogy here. The event horizon poses the same conceptual complications and philosophical implications as the afterlife, something many believe in, but no one can prove because, apart from well-known exceptions, no one has ever returned to tell us what there is after death.

Such considerations demonstrate why the idea of an event horizon is so difficult to digest for us physicists, who, through aspiration or ambition, wish to exercise knowledge that extends to every corner of the universe. For us, the event horizon is like a door that can be opened and passed through but then closes forever behind us. This is why over the years, and in particular the last decade, a considerable intellectual effort has been invested in the study of so-called black-hole *mimickers*. These are objects through which we try to mitigate some of the conceptual difficulties posed by black holes, and which we will have a closer look at in the next section.

Observations vs Experiments

We have seen that the presence of an event horizon and a physical singularity at its centre makes the very idea of a black hole difficult to digest. Given these difficulties, you might ask: 'Is it possible to do without black holes?'.

Many have grappled with this question, especially in recent years, with direct observations of objects compatible with black holes made both through the detection of gravitational waves (by the LIGO–Virgo collaboration) and thanks to the first-ever 'photograph' of a black hole (produced by the Event Horizon Telescope Collaboration (EHTC)).

Before addressing the answer, a more general consideration of how scientific knowledge progresses is necessary, especially regarding astrophysics.

Astrophysics – the only branch of physics providing evidence for the existence of black holes – is an *observational* science rather than an *experimental* one. It is impossible to carry out experiments in this branch of physics, unlike branches such as solid-state physics or elementary particle physics. Put another way, in astrophysics it is not possible to build an experimental device that allows one to produce, in a controlled manner, the physical conditions under which a specific physical phenomenon is expected to occur. By 'controlled manner', I am referring to the possibility of performing an experiment while having a clear idea of the errors that can affect the measurements. Conversely, astrophysics necessarily relies on astronomical observations to formulate theories, which can then be confirmed or refuted through additional observations. This aspect of astrophysics (which is sometimes overlooked, if not ignored) adds considerable complexity to the application of the scientific method for at least two reasons.

The first is that it is generally difficult to predict the number of sources and their properties that will be studied since only observations are made. By way of comparison, when you build a particle accelerator like that at CERN, once initial conditions have been specified (e.g., the energy of the incident beam or the properties of the target beam), you have an exact idea of the number of events or collisions that it is possible to produce in a certain period of time. Unfortunately, in astrophysics, it isn't that simple; unless you intend to repeat observations of well-known sources, the construction of a telescope is usually accompanied by a great deal of uncertainty about the number of new objects that will be observed. So, for example, when the gravitational-wave detectors of the LIGO collaboration went into operation (not forgetting that gravitational-wave astronomy is also an observational science), the uncertainty about the number of events that could be measured was of three orders of magnitude. In other words, it was not known whether there would be 400 events per year or just 0.4.

The second reason scientific progress in astrophysics is more difficult is that, by being an observational science, it suffers from a *degeneracy* on the possible explanations of the observations. Again, I will use an analogy to illustrate this concept. Imagine an old wall clock in front of you, which you can observe as much as you like. What would you notice about it? Probably that it has two hands, one long and one short, and that the shorter hand rotates 360 degrees every 12 hours, while the longer one does the same every 60 minutes. Well, these are your 'observations'. Now, suppose you have to construct a mathematically complete and physically plausible model of the internal mechanism that allows the clock to function. As you can imagine, you could find many mathematical models and combinations of mechanisms that can explain perfectly the rotation of the long hand every 60 minutes and the short hand every 12 hours. However, only one of the possible explanations is the right one, and it is even possible that you haven't considered it. Thus, to remedy this apparent difficulty, you should use all your imagination and wit in an attempt to obtain additional information. For instance, you could use microphones to listen to the noise produced by the mechanism responsible for the movement and use recordings to rule out models that are incompatible with 'acoustic observations'. Or you could measure the vibrations transmitted by the clock to the wall and work on models compatible with this additional information.

Well, progress in astrophysics is achieved in a similar way. Also, in this field, all available information is collected solely through observations, which scientists then try to explain through mathematical models built based on the laws of physics. Additional information can come from observations in different electromagnetic bands or on different observational channels, in which case we refer to *multi-wavelength* or *multi-messenger astronomy*, respectively. And, as with the wall clock, those models that appear to be more accurate and whose observations have more plausible explanations can turn out to be incorrect precisely because there is a degeneracy between the observations and the models capable of explaining them. Finally, the creativity of astrophysicists enables them to use all the possible observations (and related connections) to reduce the set of models that are in agreement with the observations or to make predictions about the behaviours that, whether observed or not, could confirm or refute the correctness of the model.

Black-Hole Mimickers

Let's get back to the idea of black-hole mimickers. Our digression into the epistemological aspects of astrophysics has been a useful illustration of how, even though the black-hole solution represents the most natural and simplest explanation to the phenomenology observed in compact astrophysical objects such as Cyg X-1, it is not the only one. There are other explanations that do not require the presence of a black hole. The characteristics of these alternative solutions are quite simple to guess. They represent extremely compact objects without an event horizon or physical singularity.[8] In fact, general relativity and several 'alternative theories' of gravity allow for a whole class of compact astrophysical objects with these properties to exist; therefore, they are dubbed black-hole 'mimickers'. Indeed, their properties so closely resemble those of black holes that sometimes they are almost indistinguishable from them, at least in terms of measurements made by distant observers such as us.

Let's look in detail at some of these mimickers. First, the so-called *gravitational star*, or simply *gravastar*. This is a solution of Einstein equations in spherical symmetry (perfectly compatible, therefore, with what we know about general relativity) first proposed in 2000 by physicists Pawel Mazur and Emil Mottola.[9] The solution supposes the existence of a layered star with an inner core (essentially the majority of the star) characterised by matter with quite unusual properties; unlike everything we are used to, it has negative pressure (all the matter we know of has positive pressure). Moreover, this core is surrounded by a very thin 'skin' of matter with equally extreme properties: an extra-strong but positive pressure.

Negative-pressure matter can be considered as a small universe that wants to expand and is associated with a 'de Sitter spacetime', after the Dutch scientist Willem de Sitter, who first proposed it. Indeed, we believe this solution of the Einstein equations is connected to the accelerated expansion we are now measuring in cosmology. At the same time, however, the thin strip of matter at very high pressure at the surface of a gravastar wants to collapse, counteracting the expansion of the nucleus and generating a stable equilibrium. In other words, a gravastar is like a balloon that has negative internal pressure but a positive outward gradient! And in which the thin outer surface prevents the core from expanding.

The most important peculiarity of this solution is that it can be constructed in such a way as to have a compactness very close to the limit of a black hole, only slightly below it, i.e., $M/R \lesssim 1/2$. In particular, it is possible to build (as mathematical solutions at least) gravastars whose surface is just above where the event horizon of a black hole with the same mass would be, that is, with a radius that is only infinitesimally larger than that of a Schwarzschild black hole. As a result, gravastars have a rigid surface just like neutron stars (hence the name, gravitational 'star'), but do not have any event horizon or contain any physical singularity at $r = 0$; thus, they are perfectly regular solutions.

The second example of a black-hole mimicker has become almost 'classic': a *wormhole*. As with *black hole*, its evocative name was also coined by John Wheeler (in an article written with Charles Misner in 1957[10]).

Once again, this is a perfectly plausible solution of the Einstein equations for an object in a vacuum and extremely compact, but whose compactness is slightly less than that of a black hole. Interestingly, because of the large curvature produced by wormholes, it is in principle possible to join two regions of spacetime that are extremely distant and asymptotically flat. In essence, the wormhole solution can be seen as a tunnel in the elastic fabric of spacetime, providing us with a 'shortcut' to connect two physically distant regions. Imagine taking a section of a flat spacetime in two dimensions (our beloved sheet in Chapter 3) and considering two points placed at a great distance from each other. If you fold one end of the sheet to obtain two parallel layers, the two points would no longer be so far from each other, which would be sufficient to be able to travel *through* the sheet instead of along it. Now, imagine a source of gravity intense enough to distort the sheet, creating a double throat, and that's it: you have created a shortcut between the two points that were far away from each other and are now very close. It has long been shown that wormholes, at least in the simplest models, are dynamically unstable; if opened, they would tend to close extremely quickly. At the same time, there are wormhole solutions that can be stabilised by the presence in the throat of exotic matter, that is, with negative mass or energy.

The schematic diagram of a wormhole reported in Figure 6.3 shows the typical 'throat' we are used to in the presence of strong curvature. In this case, however, the two throats are exactly symmetrical and can thus join two regions of spacetime that are flat at great distances. Nevertheless, it is important to note that, even in this case, the smallest section of the throat is slightly larger than a Schwarzschild black hole with the same mass would be (as shown by the dashed white circle). Therefore, this is not a black hole but simply a very compact object.

A gravastar and a wormhole differ significantly in at least one respect: the former has a solid surface, like a neutron star, while the latter has no solid surface and, at least in the simplest models, is actually a solution in a vacuum. At the same time, there are at least two properties that a gravastar and a wormhole have in common. First, both have an extremely high compactness, M/R, very close to that of a black hole, 1/2, but also slightly less. The second property they have in common is their 'genesis'. In both cases, we only have a vague idea about how these objects could be produced in nature. In other words, although gravastars and wormholes are permissible and perfectly legitimate solutions of the Einstein equations, the astrophysical processes that could lead to the formation of these objects are not known, at least at present. The situation is very different for equally compact objects, such as neutron stars or black holes. We have

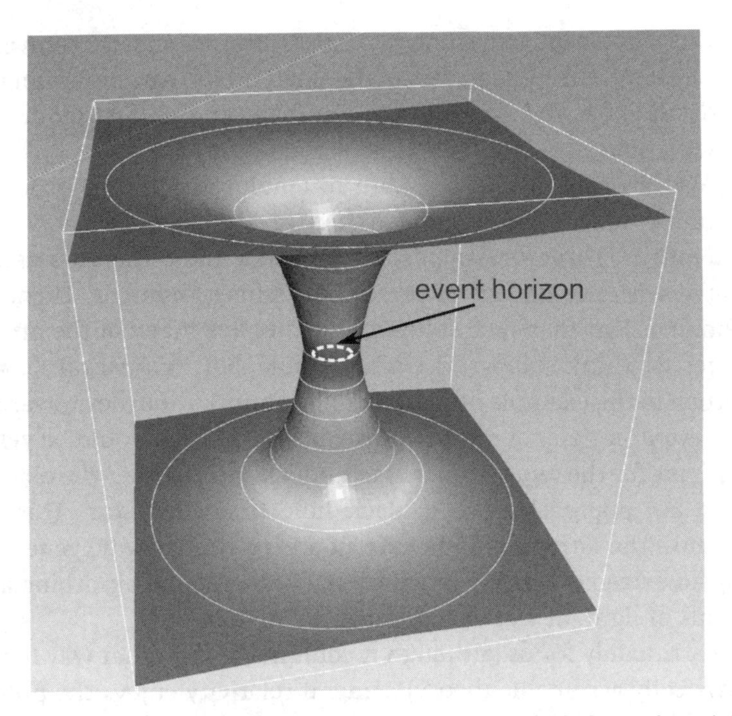

Fig. 6.3 Wormholes with two identical and symmetrical 'throats' can join two regions of spacetime that are flat at great distances. The smaller section of the throat is slightly larger than that of a Schwarzschild black hole with the same mass. A black and white version of this figure will appear in some formats. For the colour version, please refer to the plate section.

identified precise evolutionary tracks allowing us to start from stars between 10 and 100 solar masses and to arrive, through a detailed evolution of nuclear astrophysics, at a supernova and, therefore, to the formation of a neutron star or a black hole depending on the mass of the parent star. However, even using the knowledge we have today of nuclear astrophysics and the microphysical processes that can occur in a stellar context, it is not clear how it would be possible to produce a gravastar or a wormhole in practice.

In an astrophysical scenario capable of generating these types of solutions, perhaps the most serious complication is that both require the presence of exotic matter. The properties of this matter (negative energy or pressure) manifestly contrast with those that characterise the matter we can experiment with on Earth and access in the universe through astronomical observations.

Also, in this case, if we were to employ the severe logic of Occam's razor, we would have to regard such solutions as unrealistic. However, suppose we put aside the difficulties associated with the genesis of these objects for a moment. Then, it remains clear that such solutions would have many of the properties we think characterise a black hole. For this reason, if we return to the example of Carolin and Dominik's journey towards the event horizon of a supermassive black hole, it would be very difficult for the two astronauts to understand whether the object they are approaching is a black hole or a gravastar. This is because the latter would behave in a very similar way, generating an extremely strong gravitational red-shift and absorbing all forms of light, at least in appearance.

Fortunately for us (although it couldn't be any other way for a physically acceptable theory), general relativity enjoys the property of *uniqueness* in its solutions. In other words, there cannot be two solutions of the Einstein equations that are different but have identical properties. Although very similar, they must necessarily differ in something and, therefore, possess at least one characteristic that makes them distinguishable. For this reason (and thus to eliminate what would otherwise be a disconcerting degeneracy), over the last few years, scientists have studied in detail not only black-hole mimickers but principally the aspects of these objects that make them distinguishable from actual black holes.

I, too, have been involved in research of this type and have shown, for example, that a gravastar can be essentially indistinguishable from a black hole as long as we test it using electromagnetic radiation. However, it is easy to tell a gravastar from a black hole if we employ gravitational waves. Although a gravastar and a black hole are similar in size, they have a radically different interior; a gravastar is filled with exotic matter, while a black hole is empty. Thus, once perturbed, the two objects will produce gravitational waves with very different properties (we return to this subject in Chapter 7, when we consider another exotic compact object).

It is now helpful to return to the example of the clock hanging on the wall. We saw, in that case, how different models of internal mechanisms could be excluded because they were inconsistent with the observations. Similarly, the task of modern astrophysics is to produce models that use all the available information to explain the observations and rule out alternative explanations that do not correspond with these observations. Furthermore, approaches have been developed in recent years that are 'agnostic'. Therefore, they do not necessarily assume that the observed phenomenology is a black hole, leaving room for alternatives such as gravastars, wormholes or any other black-hole mimicker.

To conclude this chapter, it is important to remark on a very important concept. It is true that a degeneracy exists in the explanation of the astronomical phenomenology of objects like Cyg X-1. The alternative solutions have the very important advantage of not suffering from any of the problems inherent with black holes (such as event horizons or physical singularities). However, all things considered, the black-hole solution remains the simplest and most natural explanation within the theory of general relativity, which so far has always proved to be correct.

In addition (and this is a huge advantage over any alternative scenario), the genesis of a black hole represents the natural conclusion of the sequence of events that we know characterise the life of massive stars. It does not require improbable assumptions or the presence of matter with properties that are radically different from those we are used to. In other words, while the black-hole solution is admittedly hard to digest, alternative solutions are in many ways even harder to accept.

7 THE FIRST IMAGE OF A BLACK HOLE

So far, we have discussed in detail the properties of black holes and the intellectual challenges these objects pose. Now it is time to present a recent scientific achievement of great importance, which I have been fortunate and privileged to participate in as one of the 13 members of the Executive Committee of the Event Horizon Telescope Collaboration (EHTC).

This chapter is dedicated to the realisation of the first image of a supermassive black hole: M87*, located at the centre of the elliptical galaxy from where it takes its name (Messier-87, or simply M87). However, much of what will be discussed here will remain perfectly true and correct also when considering the supermassive black hole at the centre of our galaxy, Sgr A*, which is shown in the left-hand side of Figure 7.1.

Unless you have spent the last few years living on another planet or inside an isolated cave, it is unlikely you haven't already laid eyes on the image on the right-hand side of Figure 7.1, released by the EHTC on April 10, 2019. It is estimated that in the first 24 hours after its release, 4.5 billion people viewed the image, which has since become almost iconic and an inexhaustible source of comment and inspiration throughout social media. In less than two days, the vast majority of the human population came across this image. A record that will not be easy to beat. Even today, some years after that event, sociologists and philosophers wonder why the picture of a bright half-doughnut has attracted so much interest and become the photo of the century, worthy of a place in the Museum of Modern Art (MoMA) in New York.

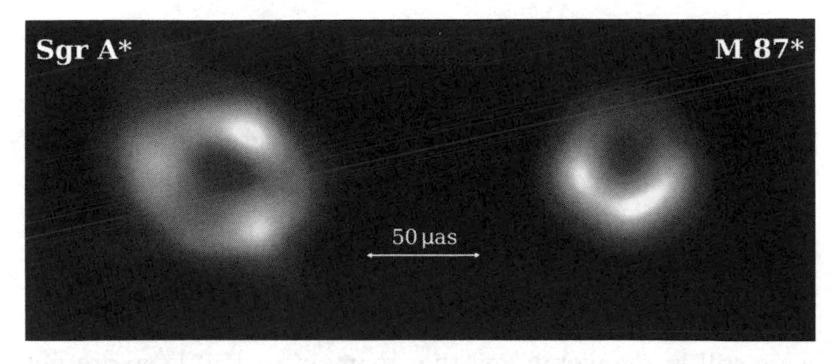

Fig. 7.1 Radio image, at a wavelength of 1.3 mm, of M87* (right) and Sgr A* (left). © The EHTC.

Given its popularity, it is possible many of you already know how this image was made and why we believe it immortalises a supermassive black hole. If not, fear not: in the following pages, I will guide you to a deeper understanding of the image, illustrating how it was produced and how the comparison between observations and theory has brought us to the conclusion that it is indeed a black hole.

However, before proceeding, it would be helpful to clear the ground of some speculations made soon after the publication of the image. It has been asserted that what is presented on the right-hand side of Figure 7.1 is not a real 'photo' but rather a map of the radio emission by M87*.

Personally, I find such considerations rather pedantic. To be clear, claims that Figure 7.1 are radio maps transformed into images are absolutely right. However, let's not forget that every photo results from a mapping process. This is true for antique photographic plates, where the image was reconstructed through the different light intensities to which an emulsion of silver salts was exposed. And it is also true for the selfies taken with smartphones, where the intensity of the electric field recorded by a sensor placed behind the phone lens transforms into a photo of us outside the Colosseum. It is even true for what you see right now while reading this text. The image you perceive of the world is only a mapping of the energy distribution of photons hitting the photosensitive cells located in the back of your cornea, being transmitted through a dense network of nerves and then interpreted by the brain.

Clearly, the point I want to make is that there is no such thing as a 'real' photo; everything we see is the result of the mapping of signals received and processed into an image. And although it is undeniable that our eyes are not sensitive to the radio band, this does not mean we cannot take a photo from this radiation. It would be like saying that the images of our children when they were still in the womb and that many of us still keep are not 'photos', just because our eyes are not sensitive to the radiation in the ultrasound band. I hope that the absurdity of this claim is evident to everyone.

To conclude, for all practical purposes, the picture on the right-hand side of Figure 7.1 is the first photo of a supermassive black hole, as produced by the EHTC. The fact that it was made starting from a radio measurement at frequencies we humans are not sensitive to is a completely irrelevant detail.

Doing What Was Thought to Be Impossible

Let me start this section by addressing a question that is as obvious as it is common: 'How is it possible to photograph an event horizon if, by definition, it cannot emit light?'. The answer is simple and categorical: you can't! Photographing an object means collecting the photons it emits, but the event horizon, by definition, does not emit any photons, and so it is 'invisible' for an external observer. That said, it is possible to take a picture of what is near and around an event horizon, especially if the horizon is surrounded by hot plasma falling onto it and emitting radiation.[1] That is exactly what we did with the telescopes of the EHTC and is what astrophysicists are interested in.

Obviously, taking a picture of what is near a black hole does not prove the existence of a black hole. However, suppose the observed image coincides with theoretical predictions that can be made based on plasma dynamics near an event horizon.

Fig. 5.1 Artistic impression of the accretion of matter in Sco X-1. The compact primary star strips matter from its companion, which is larger in size but less massive. The matter stolen in this way forms an accretion disc that slowly pours matter onto the neutron star. © NASA/CXC/M. Weiss.

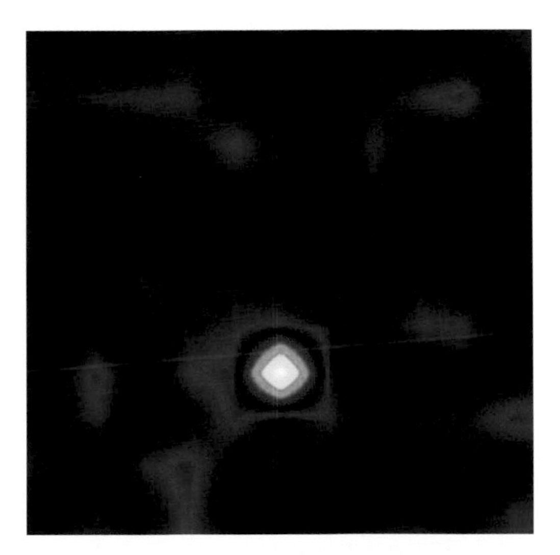

Fig. 5.2 Image in the X-ray band of Cyg X-1. The emission is linked to an accretion disc that extracts matter from a 40-solar-mass star and dumps it onto a very compact object of 15 solar masses. © NASA/Marshall Space Flight Center.

Fig. 5.3 Superposition of different images in the visible part of the spectrum of the 'Crab nebula', the remnant of the supernova explosion that took place in 1054 in the constellation of Cancer. © The Liverpool Telescope.

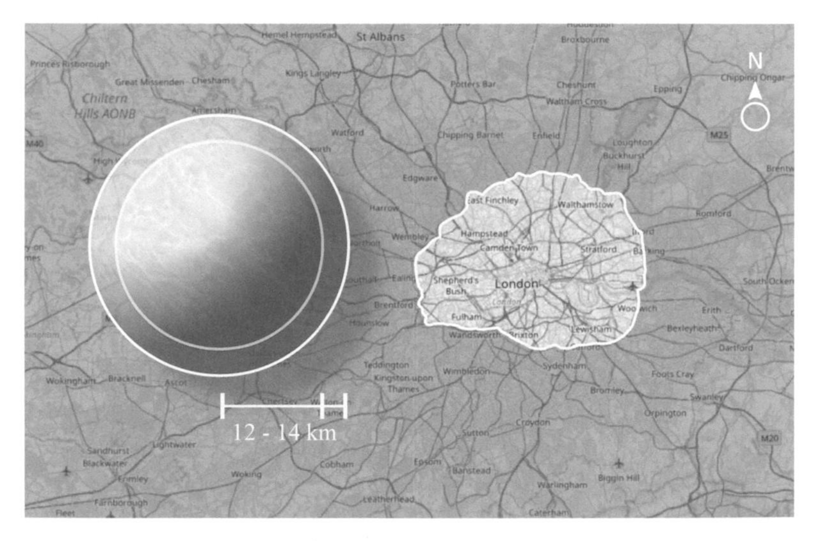

Fig. 5.4 The illustration compares the typical dimensions of a neutron star and those of the inner part of London, bounded by its north and south circular roads. © L. Weih/LR/map. OpenStreetMap contributors (CC-BY-SA 2.0)/R. K. Lazenby.

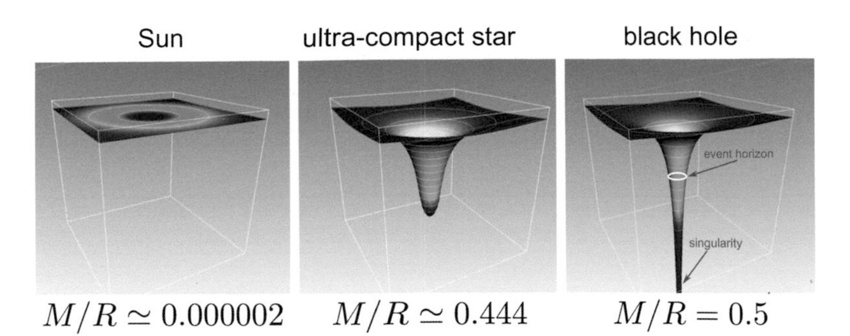

Fig. 6.1 The curvature of spacetime produced by the Sun, on the left, is created when we reduce its radius while maintaining its mass. The square in the centre depicts the Buchdahl limit while, in the one on the right, its radius has been brought to just 3 kilometres, generating a black hole.

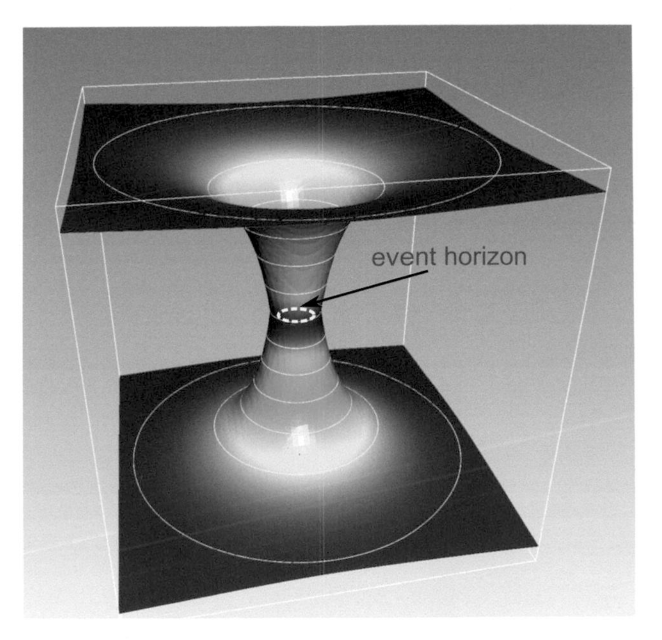

Fig. 6.3 Wormholes with two identical and symmetrical 'throats' can join two regions of spacetime that are flat at great distances. The smaller section of the throat is slightly larger than that of a Schwarzschild black hole with the same mass.

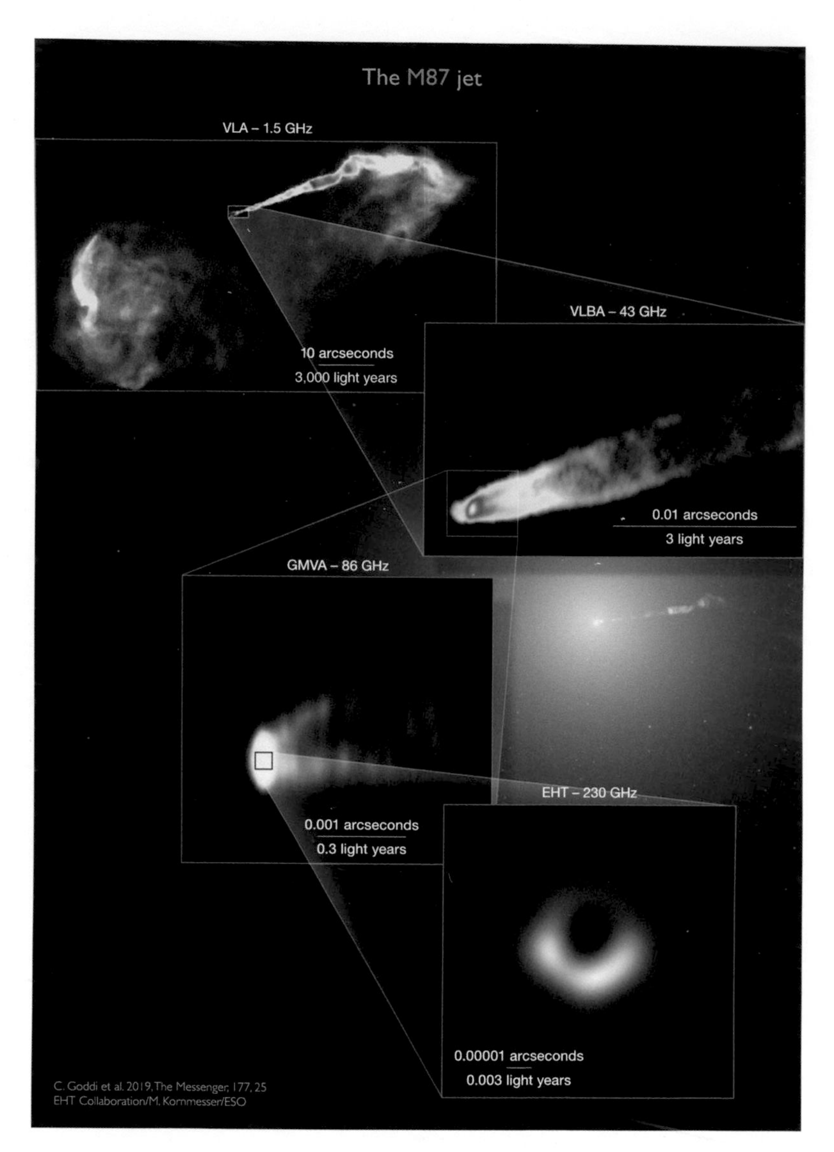

Fig. 7.2 Image composition of the M87 jet at different radio frequencies, and therefore at different angular resolutions. © The EHTC/ESO/M. Kornmesser/C. Goddi/LR.

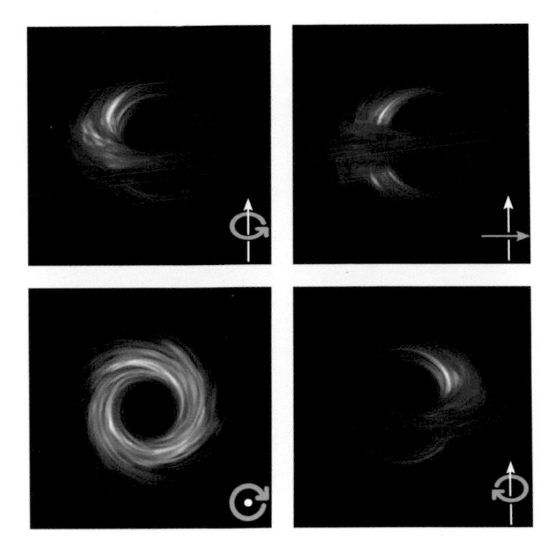

Fig. 7.9 Different-angle images of a thick accretion disc around a rotating black hole. © Z. Younsi/LR.

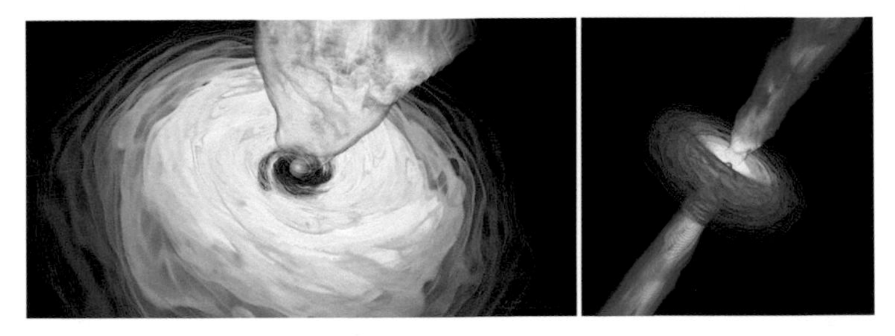

Fig. 7.10 Snapshots from a numerical simulation of the dynamics of the plasma accreting onto a black hole. Note the presence of an accretion disc of matter and of two relativistic jets. © EHTC/L. Weih /LR.

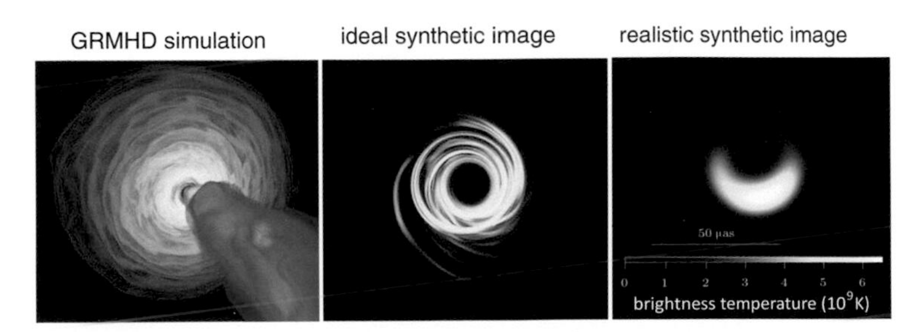

Fig. 7.11 The three phases of making a synthetic image. © L. Weih/C. Fromm/Z. Younsi/LR.

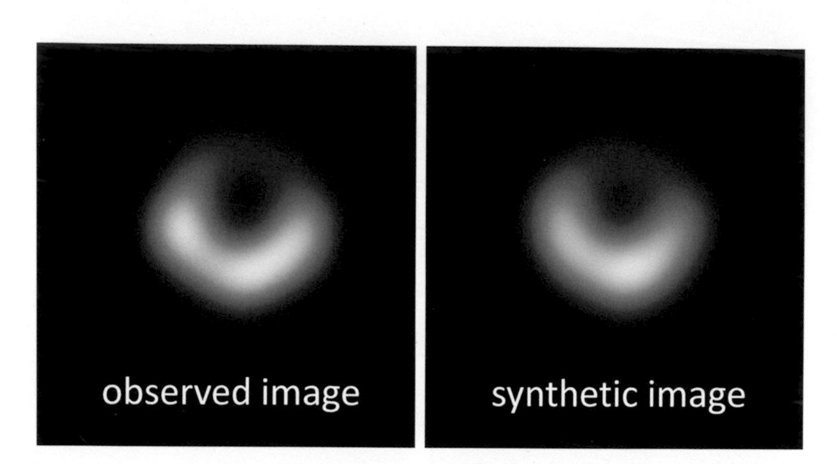

Fig. 7.14 Example of the excellent match between an observed image of M87* and a suitably deteriorated synthetic image. © The EHTC/C. Fromm/LR.

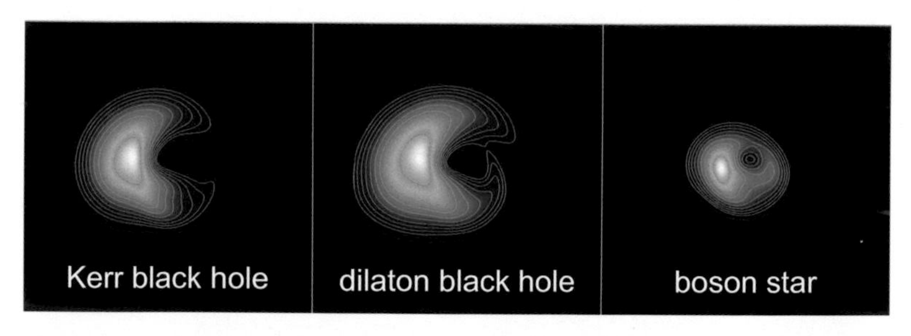

Fig. 7.15 Comparison of the images at 230 GHz of three compact objects with the same mass and subject to the same accretion processes. From left to right: a Kerr black hole, a dilaton black hole and a boson star. © C. Fromm/Y. Mizuno/H. Olivares/Z. Younsi/LR.

Fig. 8.1 Numerical simulation of gravitational waves emitted by a perturbed black hole. © R. Kaehler/LR.

Fig. 8.2 Quadrupolar deformation of the *Vitruvian Man* crossed by a gravitational wave. © M. Pössel (AEI)/LR.

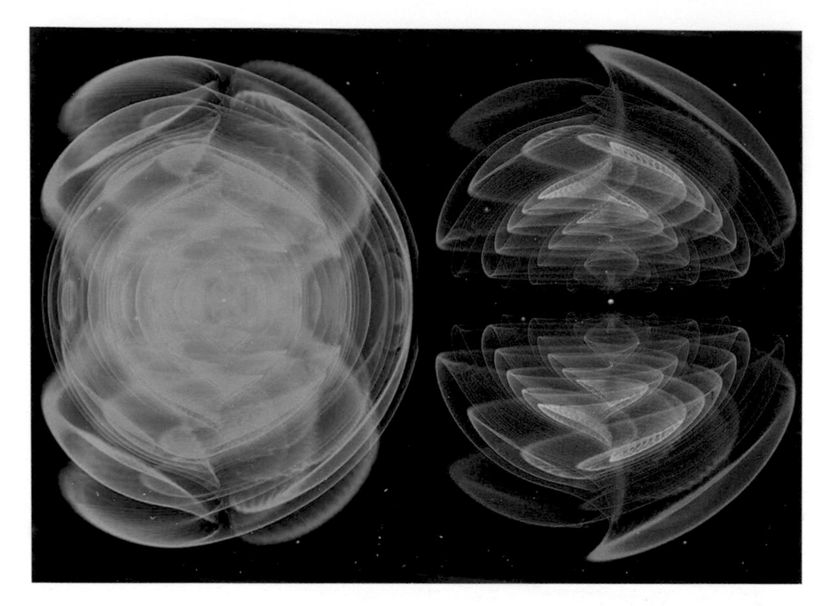

Fig. 8.6 Gravitational waves produced by the merger of a binary system of black holes. © M. Koppitz/C. Reisswig/LR.

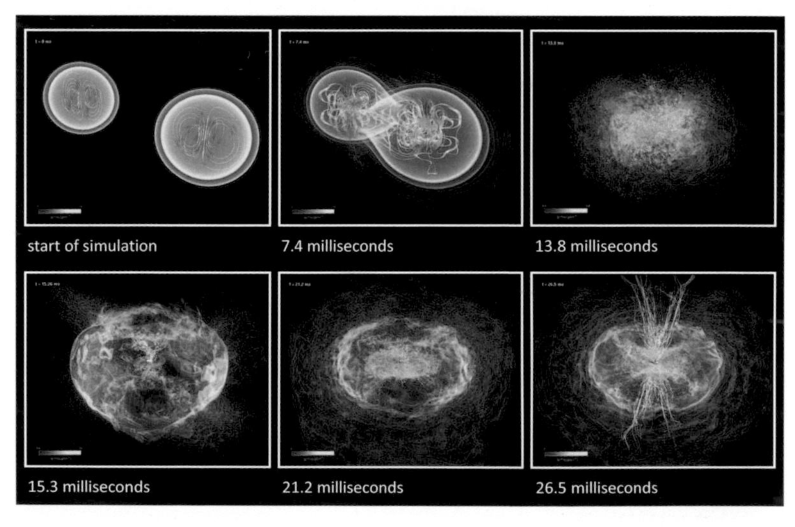

Fig. 8.7 Different phases of a numerical simulation of the coalescence of two magnetised neutron stars. © M. Koppitz/C. Reisswig/LR.

This then provides convincing evidence that a black hole is indeed present. Put differently, the picture taken coincides with the image that a supermassive black hole would produce if we got close enough.

These considerations inevitably lead to another question: 'If it is impossible to take a picture of the event horizon, but we can image what is around it, why hasn't this been done before?' The answer to this question can be deduced from what we have learned about the properties of black holes. We have seen that they are the most compact objects in the universe, so much so that if we wanted to produce one starting from the Sun, we would have to reduce the radius of our star to no more than 3 kilometres. Furthermore, it is easy to understand how difficult it can be to observe a celestial object with a radius of only 3 kilometres at astronomical distances from us, that is, well outside the solar system.

Setting aside the fact that only a tiny part of the emission from a black hole at astronomical distances can reach us, the most serious problem in making observations of black holes is angular resolution. That is, the possibility of using a telescope (or an ensemble of telescopes) sharp enough for us to observe an object of very small size in the sky. To give some numbers, a stellar-mass black hole very close to us at, say, a distance of one million astronomical units (or AU), equal to one million times that between the Earth and the Sun, would have an angular dimension projected on the sky of 10^{-10}arcseconds, or 0.0001 microarcseconds (μas), that is, approximately 3×10^{-14} degrees. By way of illustration, this resolution would allow someone in Rome to distinguish an object the size of a grain of rice in New York. There is simply no telescope, or ensemble of telescopes, capable of achieving such an angular resolution, even using the most advanced technology. As we will see in the following pages, the best resolution we can hope to achieve is in the order of tens of microarcseconds, and even that would be a technological feat.

Given these considerations, it is clear that if we wanted to take a photo of a black hole, we would need to discard stellar black holes as, obviously, they are too small to be seen at astronomical distances. Therefore, since size grows with mass (remember the expression of the Schwarzschild radius ...), we are left with supermassive black holes. The more massive the better, so they naturally appear larger at any given distance. Yes, distance also plays a role here because, as for any object, the more distant the black hole is, the smaller it will appear in the sky, even if it has an enormous mass. Hence, the 'requirements' for the optimal candidates in this photo contest are that they must be enormously massive and also close enough!

Once these two 'selection filters' have been imposed among the hundreds of thousands of supermassive black holes we presume to know in the universe, only two interesting candidates survive. These are black holes where a projected size of about 10 microarcseconds corresponds to a physical size of about 10 Schwarzschild radii.

The two candidates left in this contest are Sagittarius A* (or, more simply, Sgr A*), the black hole at the centre of our galaxy, and M87*. The first is interesting because, as can be guessed, it is the least distant from us; therefore, what it lacks in mass, it compensates for with proximity (it has a mass of 'only' a few million solar masses). On the other hand, the second candidate is more distant from us (a thousand times more, in fact) but compensates for this with mass: it is almost a thousand times more massive than the Sgr A* and among the most massive black holes known to us.

Now that we have selected our ideal candidates, we had better go hunting for these black holes ....

A Telescope as Big as the Earth

To understand how the images in Figure 7.1 were made, we first need to discuss the technical and technological aspects of the observations. For this reason, and in the spirit of inducing an appreciation of how mathematics is at the origin of all physical understanding, let me show you the following conceptual equation, which is easy to understand and very general:

$$\left(\begin{array}{c}\textit{angular resolution of}\\ \textit{astronomical image}\end{array}\right) = \left(\frac{\textit{wavelength of observations}}{\textit{telescope size}}\right) \tag{7.1}$$

Equation (7.1) tells us that if we want to achieve a very small angular resolution (so we can distinguish tiny details and study objects whose projected size on the sky is very small), we must ensure the ratio between the wavelength at which we make the observations and the size of the telescope is as small as possible.

You may think the simplest way to reduce this ratio would be to make observations at the shortest possible wavelengths, namely, those in the X-ray or even gamma-ray band. But unfortunately, this is not an option; the simple reason is that we do not receive, and cannot receive, X-ray or gamma-ray radiation from a supermassive black hole. To understand why this is the case, we need to appreciate the difference between the radiation emitted and the radiation that reaches our telescopes.

A supermassive black hole accreting matter from its surroundings will emit electromagnetic radiation in a very wide spectrum. Also, as we have seen for Cyg X-1, this radiation is extremely strong at wavelengths in the X-ray and gamma-ray bands. Such emissions are generated very close to the event horizon – precisely the region we are most interested in regarding our observations. However, the emission of radiation is a necessary but not sufficient condition for a successful astronomical observation, as we also need the radiation to reach our telescopes. Unfortunately, this is not the case with the X-ray and gamma-ray (and optical) bands. The reason is that there is quite a bit of matter between us and the source, either in the form of accreting plasma under extreme conditions of temperature and energy or in the form of simple interstellar dust. In both cases, this matter is very effective in absorbing the radiation before it can reach us. As a result, although the accreting material around a supermassive black hole does indeed emit light over a very broad spectrum, much of it is absorbed or scattered, never reaching our telescopes.

To help understand the meaning of absorption and scattering, let me offer an analogy. Take the sky on a cloudy day: if the cloud cover is not particularly dense and there is only a thin veil of clouds, we will still be able to see the silhouette of the Sun, which will appear as a brighter spot in the sky. However, if the cloud cover is very dense, then our favourite star will not be visible because the clouds will absorb the radiation it emits, so it will not reach us, or at least not directly. But, in reality, sunlight on a cloudy day does reach us (indeed, a cloudy day is not as dark as night). This is because the photons emitted by the Sun are scattered countless times by the water vapour molecules in the clouds, giving rise to the typical shadowless brightness of a gloomy day.

Let's go back now to Equation (7.1). To decide which wavelength to use for our observations and thus establish which resolution we can hope to obtain, we have to resort not to the smallest wavelength emitted but to the smallest that can be received on Earth. Omitting details concerning a comprehensive model of the radiation emitted near a black hole, the optimal wavelength for imaging a supermassive black hole turns out to be in the *radio band* and is of the order of 1.3 millimetres, thus corresponding to a frequency of about 230 GHz. This wavelength is a million times longer than that of X-ray radiation (which is around one nanometre), but it has the advantage that the plasma around the black hole is almost transparent to this radiation, which can therefore reach us without being absorbed or scattered along its path.

Having established that we need to collect radiation in the radio band, we must now tackle a much more difficult problem: we need to collect it! In particular, we have to find a radio-telescope large enough to allow us to obtain a resolution of a few tens of arc microseconds. Fortunately, radio-telescopes of considerable size can be built, as they are basically just steel structures similar to oversized satellite TV dishes. While most radio-telescopes have diameters of between 20 and 50 metres, there are two that are decidedly larger: the radio-telescope at the Arecibo Observatory in Puerto Rico and the Five-hundred-meter Aperture Spherical radio Telescope (FAST) in China, whose diameter (as the name implies) is 500 metres across. Unfortunately, however, not even such an enormous radio-telescope would be able to obtain the resolution we need, failing by at least five orders of magnitude! In fact, if you were to use Equation (7.1) to determine the size of radio-telescope required to reach a resolution of about ten microarcseconds at a wavelength of one millimetre, you would find that you need a telescope of about 10,000 kilometres in diameter! In other words, the telescope would have to be as large as the entire planet!

While this may appear to be an impossible feat, human ingenuity has allowed us to use the whole planet as a telescope by developing, in the 1980s, a radio-astronomical technique now widely used: Very Long Baseline Interferometry, or VLBI.

A Most Amazing Technique

We can avoid going into the details of the VLBI technique here, as some are rather sophisticated. Instead, it is sufficient to recall that interferometry exploits the property of 'linear superposition' of wave phenomena, whether they are electromagnetic, sound or seismic waves. This property allows waves to be combined in what is known as interference. This interference can then be constructive when the amplitude of the waves adds up or destructive in the opposite case. The VLBI technique is designed to exploit the interference between the signals received by two radio-telescopes placed thousands of kilometres away from each other, such as the radio-telescope at Pico Veleta in Spain and the Atacama Large Millimeter Array (ALMA) in Chile. In this way, it is as if a single *virtual* radio-telescope has been constructed with a diameter equal to the distance between the two *real* radio-telescopes.

The VLBI technique is truly amazing, to the point of appearing almost too good to be true, and thus raising the doubt: 'Where's the catch?'. Of course, in reality, there is no catch, but it should be recognised that the VLBI technique comes with precise constraints, some of which are rather tight.

The first such constraint is to guarantee that the two radio-telescopes observe the same object at the same instant in time.

The reason behind this requirement is quite obvious. Without perfect tuning between the two telescopes, there would be no certainty that they measure the same wavefront of electromagnetic radiation. Consequently, this would invalidate the whole interferometric process, and the observations would not lead to an image.

Let me remind you that light propagates as a wave and that the wavefronts of the light emitted by sources at astronomical distances reach the Earth as plane and parallel fronts. To ensure that two different telescopes record the same wavefront, the reading of the electric field measured by the radio receiver (even your home radio records a time-varying electric field) must be combined with the reading of the moment of arrival of the wavefront. For this purpose, each of the two telescopes must be equipped with an extremely precise atomic clock to mark the time of arrival of the electrical signal.

The second constraint of the VLBI technique is that the image cannot be produced in real time but only after correlating the signals of two (or more) telescopes. For this reason, each observation, which alone produces enormous amounts of data (of the order of hundreds of terabytes), must be recorded and sent to a special supercomputing centre, where different data streams combine to create the interferometry between the two telescopes. Further complicating matters, it is almost impossible to transfer data via the internet as radio-telescopes are located where visibility is optimal but at sites that are often isolated and difficult to reach (one of the EHTC telescopes is at the South Pole!). Notably, they are typically built in extremely arid areas – water vapour is the greatest enemy of radio observations. For this reason, once recorded on physical media, the data must be transferred (usually using students as couriers) to the supercomputing centre in charge of data combination.

The details we have discussed so far should have provided you with an initial idea of the technological complications associated with the VLBI technique and revealed, at least in part, how complex it is to employ this advanced technique in practice. Yet, with its observations, the EHTC has further raised the bar of this technological sophistication in at least two ways. The first is through the number of radio-telescopes used. The principle of the operation, previously described, between two telescopes can extend to an entire network of instruments, each connected individually to any other telescope in the network. Doing this, we gain two important advantages: the image quality is increased and the observation times are extended. When only two radio-telescopes are used, the resulting interferometric image is approximate and contains only the main basic features. Therefore, it is advantageous to use as many radio-telescopes as possible to improve this and obtain additional detail in the image. Besides obtaining more information and, therefore, detail in the image, having more radio-telescopes available also means that the source can be 'followed' over longer timescales. This is because when the source disappears below the horizon of one telescope due to the Earth's rotation, it will still be detectable by others for which the source has just appeared in the sky. For example, on any given day, the radio-telescope at Pico Veleta, in Spain, will be the first to record a source such as M87*, while the James Clerk Maxwell Telescope (JCMT) in Hawaii will be the last. Using eight different radio-telescopes distributed across the planet, the EHTC has achieved unprecedented operational continuity and conducted observations over a time frame of between 8 and 10 hours a day – five times longer than with just two radio-telescopes. In turn, this prolonged observation time has considerably improved the quality of the reconstructed images produced by the EHTC.

It should be emphasised that the simultaneous management of eight radio-telescopes distributed in remote areas of different continents is by no means a trivial operation. Especially since the EHTC does not 'own' or manage these telescopes. Instead, the EHTC obtained access to the telescopes after winning a highly competitive public application process that provides the availability of the telescopes to carry out the observations. Having obtained a common time frame for the radio-telescopes in April 2017, EHTC scientists were sent to the various sites to undertake observations – and to wish for favourable weather conditions.

While radio-telescopes can operate day and night, the weather does play a critical role because, as mentioned, water vapour considerably reduces the quality of the signal received. Although seemingly paradoxical, in 2017, EHTC scientists were trying to collect light coming from 65 million light years away (light that had taken 65 million years to reach us), and they had to hope that a cloud just a few kilometres above the ground would not absorb the radiation a fraction of a second before it reached the telescope!

The second reason why the EHTC has taken the VLBI technique to the forefront of its capabilities is its use of radio-telescopes observing at the maximum frequency of current technology, namely 230 GHz. In turn, this has guaranteed the best possible resolution to date, namely, about 10 microarcseconds (recall Equation (7.1)).

To help you appreciate the astounding 'zoom' capability achieved by the EHTC resolution in M87*, Figure 7.2 provides a composition of radio images showing how the M87 galaxy appears at different frequencies. The figure reports the radio frequency employed, the angular resolution obtained and the corresponding length-scale for each image.

As you can see, the radio-telescopes at the Very Large Array (VLA) in New Mexico operate at a frequency of 1.5 GHz and produce an image with an angular resolution of 10 arcseconds, corresponding to a scale of 3,000 light years. In this image, we can note the presence of two jets: one, clearly visible, which moves to the right, the other directed to the left and identifiable only at the end of its trajectory, when it collides with the interstellar medium (i.e., the mix of dust and rarefied gas found among stars). Instead, by increasing the observation frequency to 43 GHz, the radio-telescopes of the Very Long Baseline Array (VLBA) in the USA can reconstruct an image with a resolution a thousand times higher (namely, 0.01 arcseconds). This allows us to better visualise the interior of the jet over a length-scale of three light years. Similarly, the Global 3-mm VLBI Array (GMVA) can operate at a frequency twice as large, 86 GHz, with telescopes that are further apart, thus increasing the resolution by a

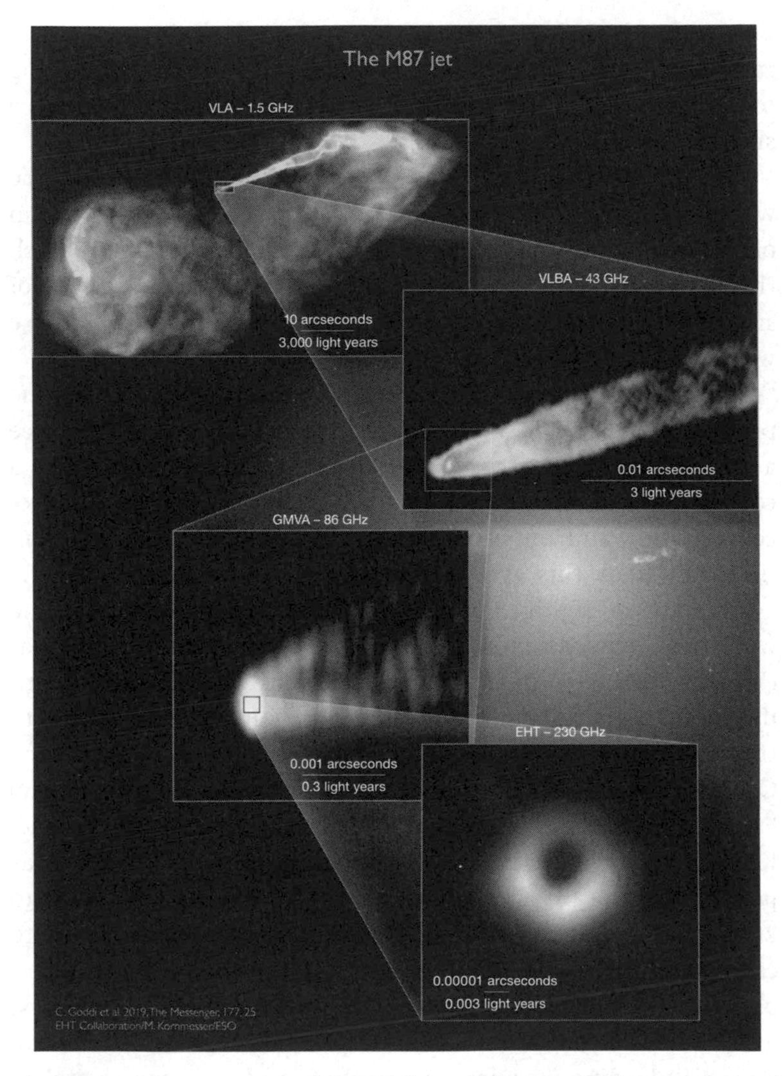

Fig. 7.2 Image composition of the M87 jet at different radio frequencies, and therefore at different angular resolutions. © The EHTC/ESO/M. Kornmesser/C. Goddi/LR. A black and white version of this figure will appear in some formats. For the colour version, please refer to the plate section.

factor of 10, namely, 0.001 arcseconds, which corresponds to a length-scale of a fraction of a light year. Finally, the enormous technological effort made by the EHTC, whose observations have been carried out at 230 GHz, have yielded an image with a resolution of the order of ten microarcseconds, corresponding to a length-scale of about 20 light hours (the event horizon has a size of only 6 light hours).

In the background of Figure 7.2, there is an image of the whole galaxy M87 taken in the optical band. It appears as an object with a diffused brightness and a bright core, from which the right moving jet seems to emerge as a collimated plume of smoke (the left moving jet is absorbed almost entirely at these wavelengths).

At this point, I think it is clear why an image like the one produced by the EHTC is the first of its kind. In order to produce and interpret the image, it was necessary to bring together a considerable number of scientists from every stage of the process, all the way from observations to theory. As in other scientific efforts of this type, success inevitably depends on the availability of significant economic resources to finance the work of so many scientists. It is not without a hint of pride that the European scientific community has provided an important, if not essential, contribution in this regard. In fact, it is thanks to the substantial funds assigned by the European Research Council (ERC) in 2013 to myself and my colleagues, Heino Falcke (University of Nijmegen in the Netherlands) and Michael Kramer (Max-Planck Institute in Bonn, Germany), that it was possible for the EHTC (whose formal foundation dates back to 2009) to 'take off' in 2014 as an effective collaboration and to carry out the much-needed research. The 14 million euros, obtained via our ERC Synergy Grant, BlackHoleCam[2] (the largest grant made by the ERC to astronomy), made it possible to set up a team of leading European researchers and has been a catalyst for the granting of additional funds for American and Asian colleagues. Only in this way, and after reaching the necessary 'critical mass', was it possible to transform the idea of taking a picture of a supermassive black hole into a reality.

I will end this section with an intriguing comment on the role luck plays in human activities, including those that we, as scientists, strive to remove as much as possible from the will of fate.

As discussed above, the EHTC has achieved the highest angular resolution ever obtained in radio astronomy for a supermassive black hole. Yet this resolution was barely enough to produce with confidence an image of M87*, the projection of which in the sky is a few tens of microarcseconds. This knowledge, however, is available to us only now that we have an estimate for the mass of M87*. Before 2017, the mass of M87* was still uncertain and the various estimates differed by a factor of two. Hence, had the mass of M87* been only a factor of two smaller (as one of the estimates claimed), its image in the sky would have been correspondingly smaller by a factor of two and the EHTC resolution insufficient, no matter how revolutionary. Because we didn't know all of this back then, the EHTC observations in 2017 were, in part, an educated guess. Nevertheless, the lesson is always the same: *fortuna iuvat audaces* (fortune favours the bold). And, in the EHTC, we were as bold as we were lucky.

All a Matter of Trajectories

Now that we have a better idea of the technology involved in making VLBI observations, we can finally turn our attention to what happens to light rays in the vicinity of a black hole. Knowing the possible trajectories light can travel on is most important in understanding why the images of M87* and Sgr A* must necessarily be as shown in Figure 7.1.

We have already discussed in Chapter 3 that nothing can move in a straight line within a curved spacetime, not even light rays (photons) insomuch that they can give rise to gravitational lensing phenomena. Since spacetime reaches its maximum curvature precisely near black holes, it is natural to expect that photons will take very curved trajectories in these regions. And that's exactly right!

We can see this, for example, in Figure 7.3, which shows the trajectories travelled by some photons near a Schwarzschild black hole, therefore a non-rotating black hole. When seen for the first time, I know that this figure can be intimidating or even confusing, but let's go through it patiently together and try to understand all of the information.

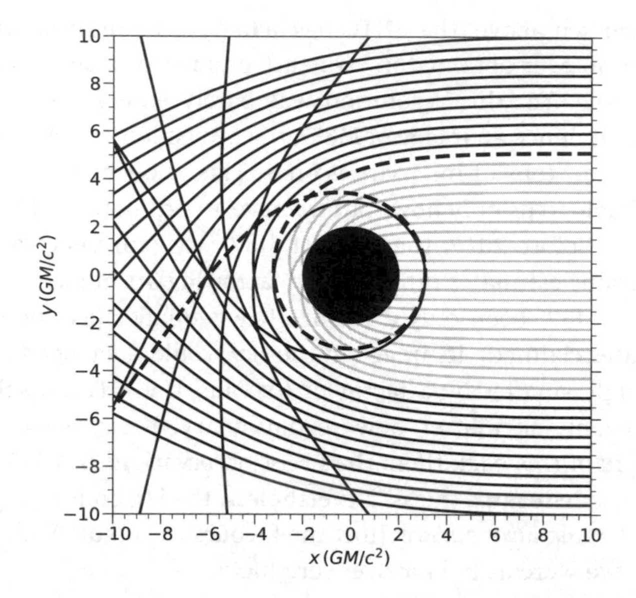

Fig. 7.3 Photon orbits on the equatorial plane, oriented from right to left, of a non-rotating black hole. © C. Fromm/LR.

To begin with, Figure 7.3 indicates the motion of photons on an equatorial plane, with the dark circle in the centre representing the event horizon of a black hole as seen from above. Furthermore, the trajectories refer to light rays emitted, all starting from the right of the figure, at $x = 10$ (GM/c^2), so you can follow any of these trajectories from their origin; for example, the top one would be at $x = 10$ (GM/c^2) and $y = 10$ (GM/c^2), and proceed from right to left.

It is fairly obvious that some trajectories are deflected but do not end up in the black hole (indicated with solid black lines), while others are captured by the black hole and end up inside the event horizon (indicated with solid grey lines). Immediately apparent is that the trajectories farthest from the black hole are deflected only slightly, but also that, as they get closer to the black hole (or, to be more precise, as the impact parameter[3] decreases), the deflection becomes larger and larger, until it is greater than 90 degrees. Note that, among the trajectories captured by the black hole, some enter 'directly' into the event horizon without undergoing a significant deflection, while others are captured after passing near the black hole. Finally, a dashed black line shows an example of the bizarre trajectories that can develop near a black hole. In this case, the photon starts from $x = 10\ (GM/c^2)$ and $y \simeq 5.2\ (GM/c^2)$, it approaches the black hole, completes an entire revolution around it, and moves away at $x = -10\ (GM/c^2)$ and $y \simeq -5.5\ (GM/c^2)$. Clearly, this is a 'strange' orbit, to say the least, but soon we will see there are even more bizarre ones.

Looking at Figure 7.3, we can imagine a limit between the black trajectories (deflected but not captured) and the grey trajectories (deflected and captured). This limit is represented by the unstable circular orbit of photons, or *light ring*, and, in the case of a Schwarzschild black hole, it is located at a precise radius[4]: $r_{po} = 3\ (GM/c^2)$. Therefore, Figure 7.3 shows us that a very special trajectory for a light ray around a non-rotating black hole must exist. Thus, if we placed, with great precision, a beam of light on this orbit, it could ideally remain on the same radius, that is, the photon ring, forever. However, this happens to be an 'unstable' orbit, and a photon moving along it would feel very much as we would if we were walking on top of a razor-sharp crest of a mountain chain with very steep sides; the slightest disturbance would make us fall to one side. Similarly, the slightest perturbation would make the photon leave the light ring and either fall into the black hole or move away to infinity.

To provide a concrete picture of what happens to photons near a black hole, I usually use a mechanical analogy based on a bow and some arrows. Imagine a target with extraordinary properties: just like a standard target, if an arrow meets it directly along its trajectory, it can be hit. At the same time, because it is a special target, it can also be hit if an arrow flies sufficiently close because the target forces the arrow to turn around (once or multiple times) until it hits it. I admit I have often wished that such a target really did exist as it would have greatly improved my performance as an archer.... However, this is pretty much what happens to all the photons grazing the black hole within the light ring: they do not need to hit the event horizon directly because, sooner or later, they will fall into it anyway. Soon we will see how the limiting trajectory given by the light ring plays a decisive role in the observations of a black hole.

The Vortical Motion near a Rotating Black Hole

The behaviour of the light rays shown in Figure 7.3 can also be extended to rotating (Kerr) black holes, which represent the vast majority of the black holes in the universe. However, before discussing the motion of light near such black holes, it is necessary to introduce an important feature of rotating black holes that we have not discussed yet. We have seen that black holes generate extreme curvature in the spacetime surrounding them. For the same reason, if they are rotating, it is natural to expect that their motion will somehow also induce a 'rotation' of the spacetime. In other words, it is plausible to expect that a rotating black hole will generate a kind of vortex of curvature in spacetime around itself, forcing any object that is close by to rotate with it.

To understand what is going on, let's imagine taking a test particle at a great distance from a rotating black hole and that this particle has no angular momentum (i.e., the particle will not be rotating with respect to a local reference system). If we drop the particle, it will naturally fall towards the black hole, following a trajectory that is, at least initially, a purely radial one. However, something bizarre happens as the particle approaches the rotating black hole. To a distant observer, its trajectory no longer appears to be purely radial, but it also acquires a spiral motion, as if it were rotating in the same direction as the black hole. Not surprisingly, if we were to repeat exactly the same experiment with a non-rotating black hole, we would find very different behaviour: the test particle would continue moving on a radial trajectory until it reached the event horizon. This phenomenon, the induced rotation of a test particle, is technically referred to as the 'dragging of inertial frames' and embodies a characteristic property of rotating black holes.

The dragging into rotation of the surrounding spacetime is not unique to black holes but also occurs with neutron stars and planets (this phenomenon has even been recorded for the Earth!). However, the strength of this behaviour again becomes extreme in the case of black holes. A rotating black hole has a very special surface located outside the event horizon, named the *ergosphere*, or *static limit*. The properties of this surface are such that any object, with or without mass, can also rotate in the opposite direction to the black hole (i.e., it can counter-rotate) as long as it is outside the ergosphere. However, once it has crossed the ergosphere, it will no longer resist the rotation of the black hole and will therefore be forced to rotate in the same direction (i.e., to co-rotate).

To better understand the role of the ergosphere, we could use an analogy from the previous chapter for a non-rotating black hole and a lake with a hole at the bottom. If you remember, we imagined a circular waterfall where the water fell along radial trajectories. Instead, we can now imagine that the hole rotates or, equivalently, that the falling water has a certain angular momentum. This would create a vortex near the waterfall where, at large distances from the hole, the current would remain essentially radial and become more and more vortical the closer it got to the hole. If we were on a boat and simply let the current carry us towards the hole, we would naturally rotate together with the rest of the water. However, at sufficiently large distances, we could use the powerful engines on our boat to resist the current and counter-rotate in relation to the hole, although doing this would become increasingly difficult as we approached the waterfall. In particular, there would be a critical point beyond which, regardless of the power of our engines, we would not be able to rotate against the current and would instead be forced to co-rotate. This critical point is the mechanical equivalent of the ergosphere in a rotating black hole.

It is essential to underline that when inside the ergosphere, we will no longer be able to oppose the sense of rotation imparted by the black hole, but we will still be outside the event horizon. So, in principle, we can still get out of the ergosphere,

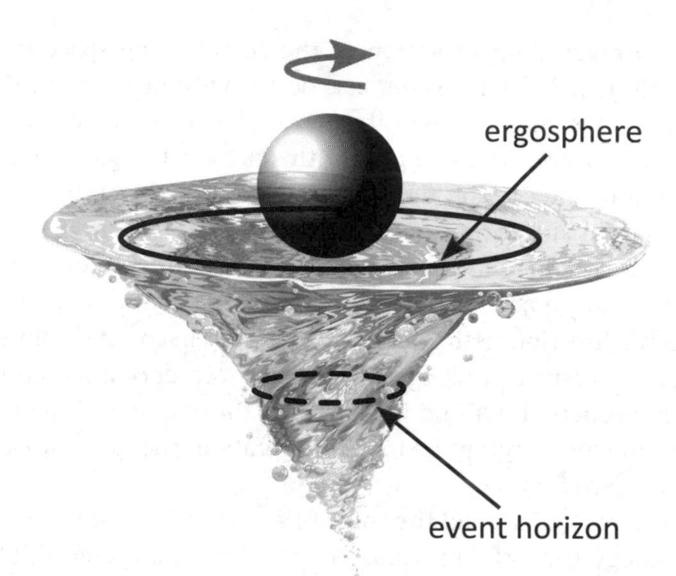

Fig. 7.4 Vortical motion of matter near a black hole and the position of the outer edge of the ergosphere, which is clearly located at a radius that is larger than that of the event horizon. © Jullands/depositphotos.com/LR.

for example, by following a partially radial motion and, therefore, moving to large distances from the black hole. This is possible because the outer limit edge of the ergosphere does not behave like an event horizon (indeed, it is at a radius larger than the event horizon), thus allowing both ingoing and outgoing motion.

Figure 7.4 illustrates this in a very schematic manner, showing the vortical motion of the spacetime near a black hole and the position of the outer edge of the ergosphere, which is clearly located at a radius larger than that of the event horizon. Outside the ergosphere, it is possible to counter-rotate. However, it is impossible to counter-rotate in the region between the ergosphere and the event horizon, but it is possible to move out to large distances. Inside the event horizon, the only motion possible is radial and inward, falling towards the physical singularity.

To give you some numbers, a Kerr black hole spinning almost at the maximum rate allowed, that is, with dimensionless spin $a \simeq 1$, the event horizon is set at a radius $r_+ \simeq GM/c^2$, while the

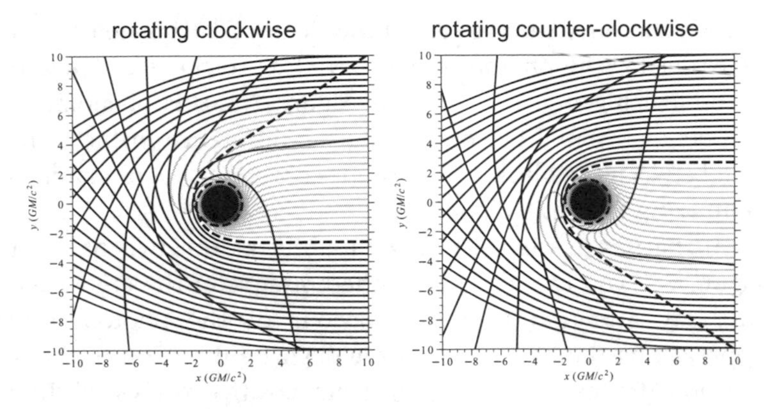

Fig. 7.5 The same orbits as in Figure 7.3, but related to rotating black holes. The rotation of the black hole creates a vorticity in the trajectories, and the photons must adapt to its direction of rotation. At the same time, the event horizon is significantly smaller. © C. Fromm/LR.

outer edge of the ergosphere is at a radius that is almost twice as large, $r_0 \simeq 2\,(GM/c^2)$.

Besides the forced co-rotation, the ergosphere is also a region near the black hole where fascinating physical processes can occur. In particular, particles can be accelerated to very high speeds, and the rotational energy of the black hole (i.e., its reducible mass) extracted with a mechanism called the 'Penrose process' (after the British physicist Roger Penrose, who first described it in the 1970s).[5] There is much that could be said on this very interesting topic, but that would move us too far from our goal....

Instead, let's go back to our light rays and examine their trajectories near a rotating black hole. Figure 7.5 describes this scenario, both in the case of a black hole rotating clockwise (left panel) and a black hole rotating counter-clockwise (right panel). In both cases, the black holes are observed from above and, therefore, along the axis of rotation.

I think it is clear that the rotation of the black hole breaks the spherical symmetry present in Figure 7.3, so the possibility of a photon being captured by the black hole depends on which side it approaches it. In particular, if it approaches from the side on which it is rotating (for example, starting from negative values of the y coordinate in the case of the clockwise black hole), it will be easier for the photon to get very close to the event horizon without being captured by it. In Figure 7.5, this phenomenon can be appreciated by considering the difference in the transition between the black and grey lines around $y = 0$, or simply by comparing Figures 7.3 and 7.5.

Furthermore, when looking carefully at the two panels in Figure 7.5, you can also notice the presence of the ergosphere. Just pay attention to the behaviour of the photon orbits approaching the black hole in the opposite direction to that of rotation (for example, starting from positive values of the y-coordinate for the clockwise black hole). You can then see that, at a certain point, the photons are forced to change their direction of motion and co-rotate with the black hole. Because the only difference between the right and left panels is in the sense of rotation of the black hole, the two images are equivalent and antisymmetric. It is possible to obtain one starting from the other by a simple rotation around the x-axis.

Note that what we have seen so far is only a small part of the general picture since our attention has been restricted to photon orbits all lying in the same plane and, in particular, in the equatorial plane of the black hole. However, the dynamics of light is much more complex if we consider arbitrary photon orbits, that is, orbits not forced to move on a precise plane. In this case, which represents the rule since only a small fraction of the possible orbits is contained in a plane, the dynamics of light rays is far more bizarre.

An example of a generic orbit, taken at random from millions I could produce, is shown in Figure 7.6. More specifically, it shows the trajectory of a photon around a black hole that is rotating clockwise. The left panel offers a three-dimensional reconstruction of the orbit, while the one on the right shows the projections of the same orbit on two planes and serves to help us understand its dynamics. Notice how the orbit is considerably more complex (but I guarantee there are even more complex ones) and how the photon can get very close to the black hole, making at least four revolutions without being

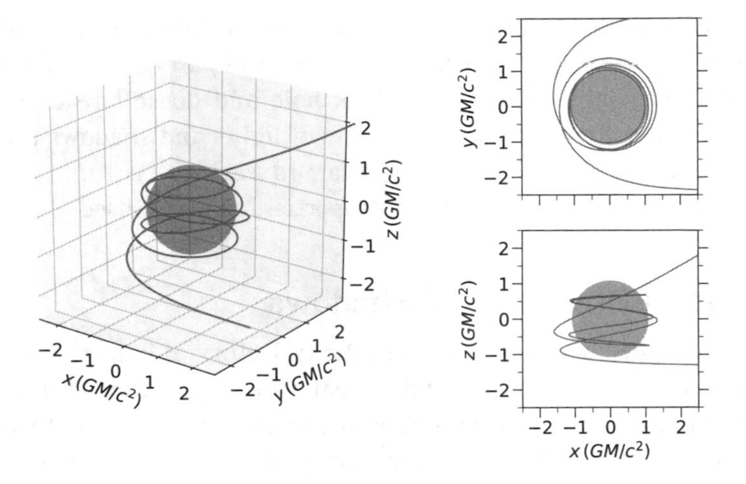

Fig. 7.6 Generic orbit of a photon around a rotating black hole. © C. Fromm/LR.

captured. In the end, it is deflected in a direction that is almost diametrically opposed to the initial direction. Furthermore, since the orbit refers to a photon not absorbed by the black hole, it is symmetric under 'time reversal'. In other words, it doesn't matter which end you start from when following it, since it can be walked both ways.

As already mentioned, the complexity of the trajectory shown in Figure 7.6 is not an exception but rather the rule. It is enough to modify the initial conditions slightly to obtain countless equally complex and intriguing variants. This property has important implications for what we will see later when I discuss how the EHTC image of M87* was 'produced' from the observations. However, if the dynamics of light near a black hole can be so complex, and a photon coming from one direction can go back in the same direction it came from (not something we are at all used to in our flat-spacetime experience!) then, reconstructing an image from all these orbits will be highly complicated. Indeed, suppose that light can be emitted and received from any direction. Then, one may even fear it is impossible to obtain a coherent photo of the regions close to a black hole or that the corresponding images will be indistinct smudges of intensity devoid of detail. However, the radiation is not emitted everywhere, only in specific regions, and part of the emitted radiation is absorbed by the black hole and doesn't reach us. This makes for a compelling display of lights and shadows that combine to produce an image that, in realistic astrophysical conditions, has very particular properties, as we will see.

A Game of Lights and Shadows

We have just seen how complex the trajectory of light rays can be in a curved spacetime and, in particular, near a black hole. We have even learnt that a photon can go back to where it was emitted. These behaviours are very different to those we are familiar with on Earth, where light essentially travels in a straight line and we see what is in front of us, certainly not what is behind us. Furthermore, if we have an object in front of us, we can receive its direct image and not an indirect image of what is behind it. Just think how surprising it would be if, approaching a man in a suit and tie, you could also see a stain on his back. Well, all this is possible in the vicinity of a black hole!

To further complicate matters, we have to consider that light not only changes direction but can also 'disappear' when captured by the black hole. To cut a long story short, it is clear that to understand what happens to light in the vicinity of a black hole, and if we want to know what to expect, we must re-educate our perception of what is possible or impossible in extreme conditions of curvature. In particular, we will now consider precise geometric shapes that correspond to potential light sources around a black hole and deduce how an observer at an astronomical distance will see them. We will thus discover that a considerable fraction of light rays can disappear to create dark areas, or 'shadows', but can also be concentrated and amplified as if through a lens.

Figure 7.7 shows the simplest scenario to consider, one that allows us to appreciate some of the subtleties of producing an image of an object in the vicinity of a black hole. This figure illustrates a non-rotating and isolated black hole (without

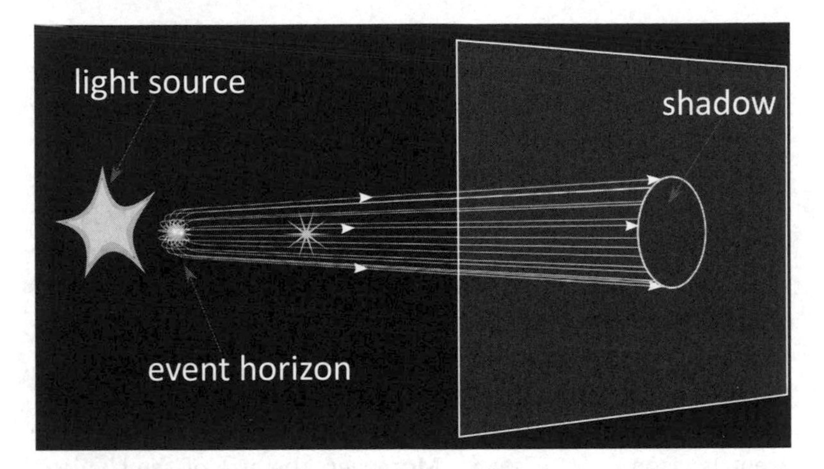

Fig. 7.7 Creation of a 'shadow' of a black hole. © T. Müller/M. Pössel (MPIA)/LR.

matter around it) whose event horizon appears as a small grey sphere. The black hole is in front of a light source, for example, a beautiful starry sky, represented very schematically in the left part of the diagram. The aim is to understand what happens to a beam of light emitted by this source and how it is received by a distant observer on the right of the image.

At this point, you should have a fairly clear idea of what happens to a beam of light shot at a black hole, and you should be able to imagine what a distant observer would see in their sky. The black hole will immediately absorb a proportion of the photons emitted by the source, that is, those whose trajectory intersects the spherical surface of the event horizon. Conversely, other photons will not intersect the event horizon but pass close enough to be deflected and captured by the black hole. The margin of the trajectories that will 'escape' the capturing power of the black hole is given by the unstable photon circular orbit, or light ring. Hence, an observer at a large distance will see an area almost devoid of light as the black hole has intercepted all the photons that should have reached them. Therefore, what will be seen, is the 'shadow' of the black hole.

The basic and yet fundamental phenomenon just described leads us to three considerations that are helpful in removing some urban legends or incorrect ideas about black-hole shadows. The first is that observing the shadow does not mean we observe the event horizon. The shadow is, in fact, determined not only by the photons that hit the event horizon directly but also by those that approach it at a distance less than the light ring. Furthermore, since it is an image projected onto the sky, it is necessarily larger even than the light ring itself. To be precise, the edge of the shadow, and therefore its size, is defined as the impact parameter of the unstable circular orbit of photons, and its radius for a non-rotating black hole is given by $r_{sh} = \sqrt{27}(GM/c^2)$. Since $\sqrt{27} \simeq 5.2$, the shadow (the darkest parts in the centres of the images of M87* and Sgr A* in Figure 7.1) is larger (by almost three times) than the event horizon $r_S = 2(GM/c^2)$. Moreover, the size of the shadow, just like the event horizon, is proportional to the mass of the black hole, so the more massive a black hole, the larger its image in the sky, and the easier it is to take a picture of it at a given distance.

The second important consideration is that the shadow represents an area in the sky where some light is 'missing', that is, the light absorbed by the black hole. However, this area does not have to be totally devoid of light. Only the event horizon cannot emit light, so it is perfectly normal that the area corresponding to the shadow also contains some radiation. For instance, a photon emitted by a source located between the event horizon and a distant observer, such as the little white star in Figure 7.7, would not be absorbed by the black hole if directed towards the observer. For this reason, the shadows in Figure 7.1 are not perfectly dark. The EHTC has actually used the contrast in the intensity between the centre of the image and the bright ring to constrain the properties of M87*.

The third and final consideration is that, with the exception of a non-rotating black hole, in which case the spherical symmetry ensures that the shadow is always the same (a perfect circle), the shape of the shadow will depend both on the spin of the black hole and on the inclination with respect to the axis of rotation at which the observations are made. This is because, as anticipated above, the rotation of the black hole generates a rotation of the entire spacetime in its vicinity (i.e., the frame-dragging effect), which allows photons that approach the black hole from one side to get closer to the event horizon than those on the opposite side (see Figure 7.5).

The presence of rotation, and thus the loss of spherical symmetry, also has another consequence: the shadow can change its appearance when viewed from different angles. In particular, the shadow of a rotating black hole will appear perfectly circular if observed along the polar direction (i.e., if we looked at it from the North Pole or the South Pole). In contrast, it would appear deformed when viewed from the equatorial plane.

Shadows and Discs

Because the shape of the shadow depends on the spin of the black hole, it is possible, at least in principle, to deduce the spin of the black hole by 'simply' measuring the shape of the shadow observed. In reality, things are much more complicated than in this simplified description and measuring the shape of the shadow with precision is still extremely difficult given the current accuracy of the observations. Essentially, it is for this reason that the observations made by the EHTC on M87* have not been able to provide a measure of its spin, even though it is improbable it is a Schwarzschild black hole. We will come back to this point when we talk about comparing the observations and predictions of the theory.

Setting aside the enormous technical difficulties that still need to be overcome to obtain a sharp image of the shadow, its shape remains a formidable tool for understanding the properties of the black hole. However subtle, all the information about mass, spin and charge (the only three properties of a black hole) are imprinted on its shadow. I find the fact we can shed light on black holes by simply measuring how they mark the absence of light as fascinating as it is paradoxical. This is true in general and applies not only to black holes in general relativity but to black holes in alternative theories and also to compact objects that aren't even black holes! We will return to these considerations when we ask ourselves whether M87* really is a black hole . . . .

Awaiting us now is an important qualitative leap in the understanding of the features that a realistic image of M87* should have. Everything discussed so far, in fact, refers to the highly idealised physical conditions of an isolated black hole in front of a background of light-emitting sources. Under realistic astrophysical conditions, however, a black hole will be surrounded by plasma at high temperatures and energies, which is falling onto it in the form of an accretion disc. So, what does a black hole look like under such conditions?

The short answer is that it will look very different from what we might expect in a flat spacetime. However, to get to the long answer, we need to proceed once again through an idealisation. So, instead of considering the image produced by a realistic accretion disc, whose properties result from sophisticated numerical simulations, let's first model the disc with a simpler geometric shape. This will help us understand some of the most salient properties of the gravitational-lensing effects (the distortions in an image due to a curved spacetime) that characterise the image of an accretion disc around a black hole.

Let's, therefore, consider an accretion disc made of a vertically thin ring of matter centred around the black hole. In other words, let's imagine putting a very thin doughnut around a black hole and asking ourselves how the doughnut would appear to us if we were to observe it from a large distance and at a certain angle.

This is just what Figure 7.8 shows, where we see, on the left, a black hole surrounded by a thin disc of matter, and, on the right, the image received by an observer placed at a certain angle of inclination. However, an important difference between this figure and Figure 7.7 is that the black hole is now not illuminated by an extended source behind it. Instead, the accretion disc itself acts as a source of light, and therefore it can appear deformed by the distortion of the trajectories of light around the black hole.

Clearly, in a flat spacetime, we would see only the portion of the disc that is closer to us, that is, the front part of the upper face of the disc, as this is the portion of the disc from which

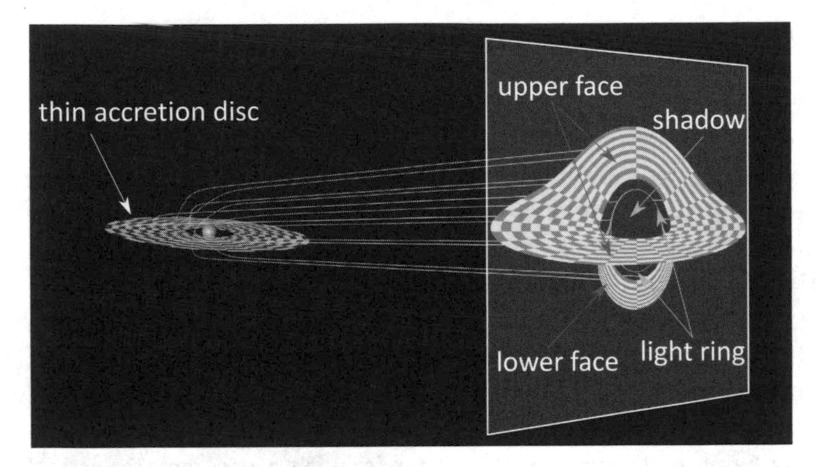

Fig. 7.8 How a very thin accretion disc would appear to a distant observer. © T. Müller/M. Pössel (MPIA)/LR.

photons moving in a straight line can reach our eyes. However, since the disc is in a region of very curved spacetime, the resulting image is rather different! In particular, Figure 7.8 shows how the 'doughnut' is very distorted. Not only would we see the front part of the upper face of the disc, but also the back part of the upper face, that is to say, the part behind the black hole and that we, therefore, wouldn't be able to see in a flat spacetime. Admittedly, this is a bizarre effect but perfectly normal in a curved spacetime. Furthermore, it teaches us an important lesson: if you find yourself in outer space and need to hide, well, don't go behind a black hole! It would be totally pointless and just reveal how unfamiliar you are with general relativity....

The bizarreness doesn't stop there. The curvature of spacetime would be such that we could also see the lower face of the back part of the disc, which would appear to us in the lower part of the image as a hanging appendage. In essence, we would see almost every part of the disc. It is important to note that the shadow would not be a single darker region but would be split into two parts: one visible above the top face of the disc and one below it. Finally, it would be possible also to observe the light ring, which in Figure 7.8 is represented by a very thin ring;

despite its thinness, it would appear relatively bright since spacetime can focus a large number of photons on this ring.

Those of you who love science-fiction movies as much as I do (especially ones with a healthy dose of realism and physical consistency) may have recognised the image of the accretion disc on the right of Figure 7.8, which closely resembles the supermassive black hole called Gargantua, which stars in Chris Nolan's movie *Interstellar* (2014). This is no coincidence; the representation of Gargantua in *Interstellar* is, in fact, quite realistic, even though it does contain at least two unrealistic aspects.

The first is that the accretion disc around Gargantua is shown as being geometrically thin, that is, with a very small vertical height. In reality, the physical conditions of the accreting plasma near a supermassive black hole lead to very high temperatures and pressures inside the accretion disc, forcing it to 'inflate', especially in the regions closer to the black hole. As a result, a realistic accretion disc is actually geometrically thick, with a non-negligible vertical height, as shown by the various images in Figure 7.9. More specifically, the four panels show four different examples of a realistic accretion disc emitting radiation at a radio frequency of 230 GHz computed from a sophisticated numerical simulation. The inclination angle of a distant observer differs in the various images, as indicated by the symbol at the bottom right of each panel, which also shows the direction of rotation of the plasma.

In particular, let's focus on the panels on the top left and the bottom right of Figure 7.9. Both represent quite generic situations and show how a realistic accretion disc would appear as more 'fluffy'. However, the basic features discussed in Figure 7.8 remain valid, namely, the possibility of seeing both faces of the disc and the doubling of the shadow.

The second important aspect, shown in the panels of Figure 7.9 and absent in the image of the disc around Gargantua, is that all discs have a side that is brighter and, consequently, a side that is less bright. The only exception, which I have decided to show precisely because it is an exception, is illustrated in the bottom left panel and refers to the case

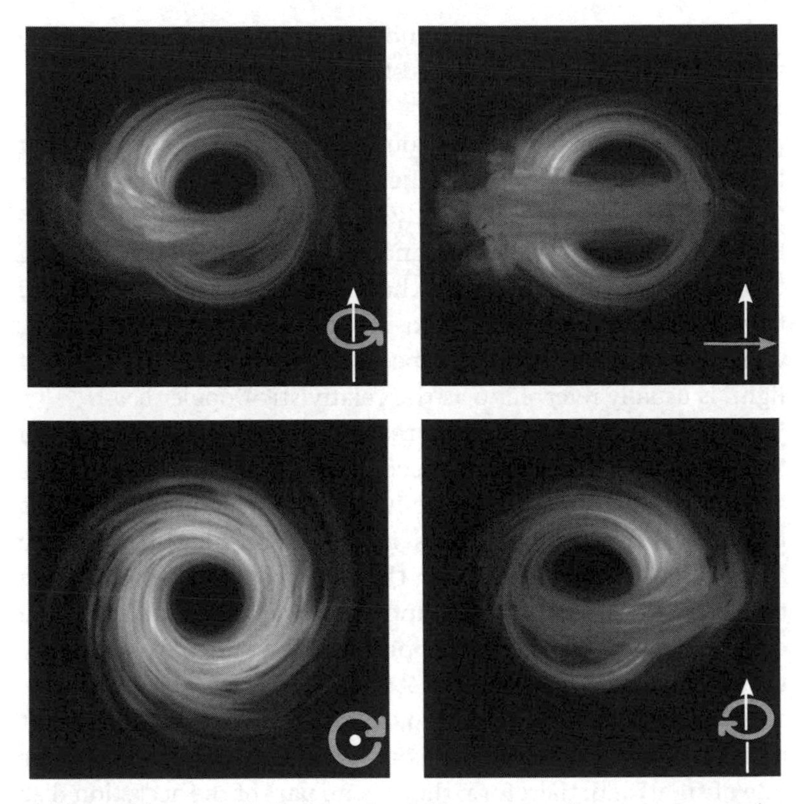

Fig. 7.9 Different-angle images of a thick accretion disc around a rotating black hole. © Z. Younsi/LR. A black and white version of this figure will appear in some formats. For the colour version, please refer to the plate section.

of a zero inclination (i.e., when viewing the accretion disc from the direction of the North Pole).

The origin of the asymmetry in the brightness between the two sides of the disc is linked to a common phenomenon, the Doppler effect (after the Austrian physicist who first described it in 1842). This effect is responsible for the variation of the frequency of a wave signal when emitted by a moving source. In other words, it describes the well-known variation in frequency of an ambulance's siren as it moves towards or away from us. We have all experienced an ambulance approaching us and the frequency of its siren increasing in what is known as a *blue-shifted* Doppler signal. For the same reason, the siren's frequency decreases as the ambulance moves away from us, producing what is described as a *red-shifted* Doppler signal.

Although the Doppler effect is already present in classical physics, and we experience it routinely for sources that move at low speeds in respect to us, it takes on new relevance in relativity, as the emission changes not only in frequency but also in amplitude. More specifically, the signal emitted by a source moving towards us at speeds close to that of light will be both shifted towards the blue and amplified in intensity. The amplification, which is particularly large for speeds of the source close to that of light, is usually referred to as the relativistic Doppler *boost*.

Returning to Figure 7.9, the reason three of the panels have a 'bright side' is that, in all three, there is matter that rotates around the black hole, so there is a side in which matter moves towards us and its brightness is, therefore, amplified by the relativistic Doppler boost.[6] For the same reason, matter on the opposite side of the disc is moving away from us, and the corresponding emission is suppressed. This also explains why the bottom left panel in Figure 7.9 does not have a bright side or a dark side. For such inclination, there is no area of the disc that moves towards us (all motion is in a plane perpendicular to our line of sight) and, therefore, there is no part of the accretion disc whose emission can be amplified in any way.

The Predictions of the Theory

We have seen how the radiation emitted in the vicinity of a black hole can be subject to incredible tricks of light and shadows. We have also seen how an object with a disc geometry (be it thick or thin) emitting radiation in the radio band may appear radically different from what we would expect of our experience in flat spacetime. What remains to be seen and understood is what the physical properties of this emitting object should be. In other words: 'What happens to matter near a supermassive black hole?'.

An easy way to answer this question would be to take a certain amount of matter, such as a large bucket of water, and throw it over a rotating black hole. We could thus observe what happens to the matter as it falls on the event horizon and use the data obtained to infer what generally happens to accreting matter on a supermassive black hole.

Sure, it would be very convenient to be able to do this, and it would solve a lot of problems for me personally, even though, in all likelihood, it would leave me unemployed. However, it is obvious, unfortunately, that this path is not viable. So, the only way (and I must emphasise the 'only' way) to answer this question is to take the equations that describe the behaviour of water poured onto a black hole and solve them with the help of powerful supercomputers. I have learned to do exactly this ever more realistically throughout my scientific career.

I am well aware that when I say I make a living solving partial differential equations involving black holes and neutron stars, I am widening the gap between myself and my interlocutors. To avoid doing this with you, I will try to explain what this amounts to by using an analogy I am sure you are familiar with. Solving differential equations to understand what happens to matter in the vicinity of a black hole is, albeit with some differences, similar to what is done to obtain weather forecasts. For many of you, a weather forecast boils down to a series of icons on a website or a geographic map with symbols representing rain, clouds or sunshine, illustrated by a presenter on television. In reality, however, those icons and symbols on the map are the results of complicated calculations made several times a day, using parallel supercomputers that solve the equations of classical viscous hydrodynamics (also known as the Navier–Stokes equations). It is no coincidence that these are also partial differential equations. In other words, the supercomputers start with fairly precise information (collected by ground-based stations or airline pilots) about the weather conditions, such as the wind speed, temperature or the density and composition of the atmosphere in different locations. These pieces of information form the 'initial conditions' and are then used to solve the Navier–Stokes equations on a given region with a given angular resolution. In this way, it is possible to obtain predictions for the velocity, temperature and density distributions of the atmosphere in the future. This whole complicated process is summarised as 'weather predictions', which somewhat trivialises it. In reality, what we obtain is the solution of beautiful and complex equations, once again evidence that mathematics pervades our world more deeply than it may appear at first sight.

Well, together with my colleagues, what I do is rather similar, at least in principle. I, too, have to start from the initial conditions (for example, the quantity of water in the bucket poured onto the black hole and its composition) and solve equations that tell me where that water goes, if it remains as it is, or whether it vaporises into an ionised plasma. The equations I am dealing with are those of general relativistic magnetohydrodynamics (which describe the dynamics of a plasma in the presence of intense magnetic fields and strong spacetime curvature). And I do not deny that they are much more complex than those used in weather forecasts to compute the cloud coverage of a particular geographical area. The physical regimes in which these equations are solved are also much more extreme, with variations in temperature and densities that are the most severe imaginable.[7]

Notwithstanding these (important) differences, and at the risk of oversimplifying things, it is possible to say that what we do in our simulations of plasma dynamics is not so different from weather forecasting. Clearly, our simulations are not aimed at predicting the weather in the vicinity of a black hole but at establishing what happens to the matter that has found itself, for various reasons, orbiting the black hole.

Thus, the simulations tell us that the plasma does not accrete onto the black hole in a spherically symmetric manner; that is to say, it does not build up a single spherical cloud in which the plasma falls onto the black hole following radial trajectories. Rather, it concentrates on a plane referred to as the 'equatorial plane', forming a disc that rotates around the black hole while accreting onto it. The simulations also reveal that the accretion disc around a supermassive black hole such as M87* cannot be geometrically thin, as portrayed in the *Interstellar* movie. Instead, it must have the shape of a thick doughnut and, moreover, a doughnut that is neither stationary nor perfectly symmetric. Numerous processes and instabilities driven by the strong magnetic fields develop in the accretion disc and are responsible for dissipation and the ultimate falling of the matter onto the black hole.

The nice thing about numerical simulations is that they give us access to all the properties of the plasma, such as density, velocity, temperature and pressure. In this way, we can calculate how much matter falls onto the black hole per unit of time (technically referred to as the 'accretion rate') or how strong the magnetic field is near the event horizon. Through the simulations, we discover, for instance, that the accretion rates of both matter and magnetic field are not steady over time but are 'quasi-stationary'; that is, they fluctuate around some average value that remains only approximately constant over time. Not very different to what happens to the water falling from a waterfall: the amount of water falling per unit of time is only approximately constant, accompanied by fluctuations associated with times in which there is either more or less water falling.

Finally, the simulations reveal that the accretion process inevitably leads to the development of 'relativistic jets' in the polar areas of the black hole. That is to say, regions where there is very little matter (almost as if there was a vacuum), but where magnetic fields are instead very strong. Indeed, magnetic fields are the dominant force in these jets and essentially determine the dynamics of the plasma. Because of the underlying symmetry in the accretion process, these jets normally come in pairs, with the magnetic field essentially aligned in a direction perpendicular to the equatorial plane of the disc. Decades-long VLBI observations of the jet in the M87 galaxy have revealed that the minimal amount of matter present in the jet essentially moves at the speed of light.

An example of the results of the simulations is shown in Figure 7.10, which reports two images from a general-relativistic magnetohydrodynamics simulation carried out by our group in Frankfurt and showing the dynamics of the plasma near an accreting rotating black hole. It can then be noted how the

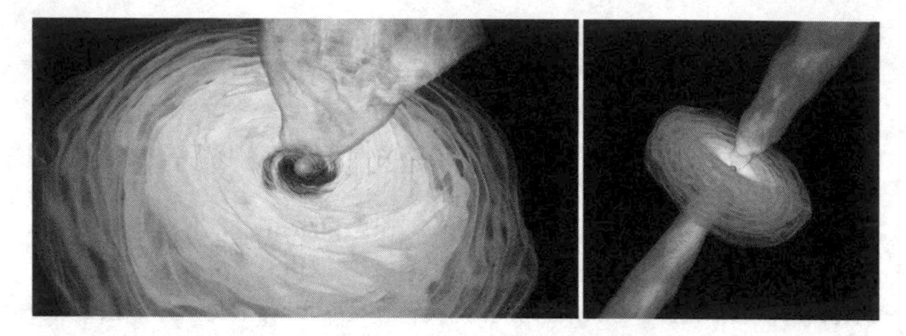

Fig. 7.10 Snapshots from a numerical simulation of the dynamics of the plasma accreting onto a black hole. Note the presence of an accretion disc of matter and of two relativistic jets. © EHTC/L. Weih /LR. A black and white version of this figure will appear in some formats. For the colour version, please refer to the plate section.

accretion disc is thick, turbulent and non-uniform, thus leading to a rate of accretion that is only quasi-stationary. Also emerging in the direction orthogonal to that of the accretion disc are two relativistic jets; these are essentially regions where there is very little matter but very strong magnetic fields and where particles can be accelerated to a very high speed.

Unfortunately, with the exception of the speed of propagation, very little is known about the composition of the matter inside the jet, although it seems plausible that it is made up of very light charged particles, such as electrons and positrons.[8] Also unknown is what launches the matter inside the relativistic jets and accelerates it to enormous speeds, thus transferring vast amounts of energy. As you can easily imagine, there is no shortage of ideas about what is behind the launching and acceleration of these jets, but also no firm conclusion as yet.[9]

Finally, to add a further veil of uncertainty to the many that already envelop relativistic jets, we do not understand well the processes behind the collimation and stability of these jets; namely, processes that allow them to retain their thin elongated shape over enormous length-scales without breaking apart or dispersing in the surrounding space. These processes are so efficient that the jet of M87 remains collimated on a length-scale that exceeds its section by eight orders of magnitude. To appreciate what this means, let's consider a relativistic jet as if it were the smoke coming from an ashtray containing a discarded cigarette with a diameter of about one centimetre. Well, to replicate what we observe in M87, the cigarette smoke would need to remain collimated, without breaking, over a length of about a thousand kilometres, more than the distance between London and Inverness, in Scotland. So, it should now be clear that, whatever its origin, the collimation process is as effective as it is extreme!

Let me end this section with a deliberation that I consider very important in the context of scientific dissemination, especially nowadays when it is so difficult to distinguish truthful information from misinformation, science and rigour from ignorance and superficiality.

It is essential to underline that there is a fundamental difference between the images created through numerical simulations and those that we often see on the glossy covers of popular-science magazines or websites dedicated to public outreach. The latter are, in general, 'artistic impressions', that is, suggestive representations of how black holes ought to appear in artists' interpretations. Such images are often the result of a collaboration between experts in computer graphics and astrophysicists, but, despite the best of intentions, they are not necessarily consistent with the laws of physics and, most often, constructed simply to please or excite our imagination.

By contrast, the images of black holes or jets obtained through numerical simulations, which may even be interesting from an 'artistic' viewpoint (see, for instance, Figure 7.10), are the result of the solutions to complex equations (examples of these equations reported in this book are of general-relativistic magnetohydrodynamics around a rotating black hole). By their very nature, therefore, these images are consistent with the laws of physics, as they are based on equations that we consider to be faithful representations of reality near a black hole.

However, such images do not yet tell us how a supermassive black hole would appear to an observer. If we want to understand what we would see if we were to observe a supermassive black hole with radio-sensitive eyes, we need to make an additional effort.

From a Turbulent Plasma to a Map of Light

Although physically consistent and possibly graphically appealing, the images resulting from numerical simulations do not reveal what we might see if we were near a rotating black hole and our eyes were sensitive to radio waves. These images tell us how *matter* is distributed and not how and where *light* is produced. Obviously, if there is no matter, there is no light source either, so there is no possibility of creating an image. However, the knowledge of how matter behaves around a black hole is, per se, not sufficient to actually produce an image of an accreting black hole. What we still need to determine is how this matter can actually shine and how the light produced is thus distributed around the black hole. This is because not all of the matter present is a source of light, and not all of the matter that can produce light does produce the light that we receive.

To make this concept slightly more understandable, think of the strings of coloured lights that we use to decorate Christmas trees. Let's imagine one consisting of a long sequence of coloured LED lights, where not all the lights are working because the last celebrations were quite lively, and somehow abundant amounts of sparkling wine managed to shower the tree . . . . Let's say that only half of the lights are working, but we do not know which are failing. We could take our string of lights and decorate our Christmas tree, noting the exact position of each light as the sequence is distributed on the tree. At this point, we could turn off the lights in the house and turn on the Christmas tree lights. As you can imagine, although we know the position of each light, we cannot yet know what 'image' of the tree we will receive, that is, what sequence of lights we will obtain once we turn them on. In particular, we do not know whether a light of a given colour and position will emit light or not, so we are left with a wide assortment of possible combinations, all of which are viable, but only one of which will actually be realised in practice when we turn the decorations on . . . .

Well, this is not so different from what happens with our simulations of the plasma dynamics around a black hole. Our solutions tell us precisely where the matter is and what its properties are. However, it does not reveal which parts of this matter can emit light, nor how intense this radiation is. So, as you can imagine, we now have to deal with an additional complication. Together with understanding what happens to matter around a black hole, we also need to introduce an *emission model*, a model for the emission of light in the accretion flow. In other words, we need a mathematical model that allows us to associate the properties of the plasma most important for the emission of light (the density and temperature of the electrons) with a physical model expressing how much radiation is emitted in the radio band. Going back to our Christmas decorations, we need a model that tells us which of the lights in the string is working, its colour and the brightness of the light it can emit.

Finally, as if the whole framework for the emission of radiation was not complicated enough, we have seen that the light emitted around a black hole can follow very bizarre trajectories. As a result, we need to ascertain whether the radiation from a region of the plasma having the correct conditions to emit such radiation would actually reach a distant observer. At the same time, we cannot discard a region of the plasma with only a weak luminosity on the assumption that this may be undetectable. It is possible that the emission from this region is weak, but it could be intensified by gravitational lensing and relativistic amplification, thus becoming visible to a distant observer.

To recap, I think it is now clear that performing simulations of accreting plasma is only the first step in obtaining the image of a supermassive black hole. The next equally challenging and complex step is to predict how the computed distribution of matter emits light and how this light propagates near the black hole until it reaches us. To accomplish this requires solving the 'radiation-transport' problem, which amounts to calculating the trajectory of each light ray in the curved spacetime around the black hole and evaluating the interaction that the light ray will have with the matter it encounters on its way. Under realistic conditions, light rays will not propagate in a vacuum, and their trajectories will inevitably intercept (often repeatedly) the accreting matter before reaching us, the observers. Furthermore, depending on the physical conditions encountered by the light ray at each point of its path, the interaction with the plasma can either decrease the intensity of the light ray (if the matter absorbs the light ray) or increase it (if additional light emitted by that portion of the plasma is added to the initial ray). Given the complexity of computing these variations, which will take place along the trajectory of each light ray for millions of rays, this step in the theoretical construction of the image can only be accomplished through expensive simulations performed on parallel supercomputers. And that's exactly what my research group in Frankfurt has done.

I am leaving aside the technical details of how the problem of radiation transport is solved in practice, which are numerous and not particularly important for this discussion. Instead, in Figure 7.11, I have schematically brought together the three most important phases of the process that ultimately lead to the construction of a *synthetic* (i.e., purely theoretical) *image* of an accreting supermassive black hole.

In particular, the left panel shows the result of a simulation of plasma dynamics, revealing the distribution of the plasma in the accretion disc and the magnetic fields in the jet (see also Figure 7.10). The panel in the middle, on the other hand, shows the emission in the radio band (particularly at the frequency of 230 GHz) as calculated based on the distributions shown in the left panel. In other words, it illustrates how the matter placed in

Fig. 7.11 The three phases of making a synthetic image. © L. Weih/C. Fromm/Z. Younsi/LR. A black and white version of this figure will appear in some formats. For the colour version, please refer to the plate section.

the left panel would appear if we had eyes sensitive to radiation in the radio band. Finally, since the central panel in Figure 7.11 shows an *ideal* synthetic image, that is, relative to a perfect instrument without any noise, the panel on the right shows how the emission would really look if recorded through a network of radio-telescopes connected with the VLBI technique and, therefore, with a resolution that is high but finite. All the images in the three panels are made with a viewing angle reflecting that at which we think we are observing M87*.

Overall, Figure 7.11 offers us a cue for several considerations. First, when moving from the left panel to the middle panel, we can readily appreciate how not all of the accreting matter contributes to the emission of radiation in the radio band (for instance, the jet contributes little in this specific image). Furthermore, the maximum intensity of the emission is not always linked to areas with maximum density, even though, in general, a good portion of the emission comes from the inner parts of the disc, which are also those where the density is highest.

Second, the image in the middle panel represents the combination of all the light contributions produced in the vicinity of the black hole and which, once deflected and amplified, can reach a distant observer. Hence, although the light we see in the picture is emitted largely by the disc, it is also produced by the jet moving towards us (i.e., the 'approaching' jet), and even by the jet moving away from us (i.e., the 'receding' jet). This is possible because, as we have seen, light emitted in the opposite direction to our line of view can still be deflected and reach us if the accreting matter in this radio band is 'optically thin', essentially transparent. However, if it were 'optically thick', the light from the receding jet would be absorbed, and the jet would be invisible.

Finally, the panel on the right reminds us that the resolution it is possible to obtain with current technology, although the highest ever, is still far from what we would like and from what would allow us to derive extensive information from the image of a black hole. However, Rome was not built in a day, and the process of producing the image of a black hole does not end with the results obtained by the EHTC in 2019. On the contrary, the image of M87* presented in Figure 7.1 must be considered as the beginning of a process that will take decades to reach a complete development.

Before concluding this section, let's return for a moment to the analogy of the matter accreted by a black hole and the flow of water measured at a waterfall. We have seen that the motion of matter around the black hole is turbulent in nature and therefore fundamentally irregular, even if it has precise statistical properties, such as average values and fluctuations of almost constant amplitude. If the motion of matter is turbulent and fluctuating, then so too must be the emission of light from that matter. The implication of this is as trivial as it is important. There cannot be a *single* and *static* image of M87*; the image must necessarily be dynamic and fluctuating to represent the state of matter around the black hole over a certain time.

When you come to think of it, this conclusion isn't surprising at all. Imagine taking a series of photos of a mountain at different times, similar to making a video with the time-lapse technique. The various images we take will be very similar, but none will be identical to the previous one, perhaps because a cloud has moved or the lighting conditions have changed slightly. What happens to our simulations, and therefore to our radio observations, is not very different when we try to produce an image of M87*. The distribution of matter will change continuously, and so will the radio emission. So, the image we are trying to produce will inevitably be subject to variability, much like taking a picture of a very active toddler who pays no heed to our appeals to stay still. In such a case, we all know that the only way to get images that are not shaky is to use exposure times that are shorter than the typical timescale over which the subject changes. In essence, we need to shoot faster than they move!

So, you may be wondering: 'What is the timescale of variation of the emission in M87*?'. Although it reflects very complex dynamics, the answer to this question is fortunately very simple: the timescale depends on the black hole mass. To be precise, it can be associated with the time needed to travel roughly a circular orbit around the black hole. In the simplest case of a non-rotating black hole and an orbit placed near the light ring, this timescale is given by $\tau \sim (3GM/c^3)$, where the estimate takes into account that it is sufficient to consider only a portion of the orbit and that the latter can also be at radii smaller than the light ring.

If we calculate this timescale for the estimated mass of M87*, that is, 6.5 billion solar masses, we would conclude that the variability timescale is of the order of 24 hours. Therefore, although the source of radiation around M87* is intrinsically fluctuating, it will effectively appear as nearly static provided we use exposure times of less than one day. Since the EHTC observations took place over a time window of about 8 to 10 hours (the timescale over which M87* becomes visible to the first telescope in France and then disappears for the last telescope in Hawaii), the VLBI images are essentially static. If a variable is present, this should be minor and only appear when comparing across different days – which is exactly what the observations have shown!

A quasi-stationarity in the emission is unfortunately not present in the case of Sgr A*, whose mass is almost a thousand times smaller than that of M87* and whose expected variability is, therefore, of the order of only a few minutes. Under these conditions, the VLBI observations would somewhat resemble a short and incomplete film, which is, of course, much more complex to analyse and interpret. After three years of additional work and analyses, in May 2022 the EHTC published the results of the observation of Sgr A* and the corresponding image is shown on the left-hand side of Figure 7.1. Once again, the image shows a bright ring of emission around the shadow. Despite the differences in masses between the two black holes (Sgr A* is a thousand times less massive than M87*) and in the physical environments (Sgr A* accretes a million times less mass than M87*), the images are remarkably similar. This is what we would expect given the properties of black holes and their simple scaling with mass discussed in Chapter 6.

I want to close this section with a question some might ask: 'Should we take into account changes in the mass of the black hole due to the accretion of matter during the observations?'.

This question would be very interesting, and I can provide a simple and categoric answer: no, we need not worry. The timescales for the variation of the mass of a supermassive black hole are, in fact, enormous compared to those over which the observations are carried out. To grasp what this means in practice, let me remind you that the observations of M87* have allowed us to estimate that its mass increases by one single solar mass over 400 years of accretion. Considering that the total mass of M87* is billions of solar masses, then over 10 million years, it will increase its mass at most by one part in a million. After all, this is not so different from when we take our time-lapse photos of the mountain. Also, in this case, we know the mountain will change while we take a series of pictures; slowly but inexorably crumbling due to erosion. So, in about 10 million years, a large part of it will end up in the sea. However, because we know these changes are so minute, we simply don't think about them when taking our time-lapse video of the beautiful mountain....

The Images of M87*

In the previous pages, I have emphasised that carrying out plasma-dynamics simulations is conceptually not dissimilar to what meteorologists do with their weather forecasts. However, there is a significant difference: while weather predictions refer to a relatively narrow window in the future, ranging from a few hours to a few days, after which they lose reliability, our simulations cover much longer timescales of the order of about a decade.

This point may be quite surprising for some of you and begs the obvious question: 'Why on earth do the simulations cover a period of 10 years when the observations last less than a day?'. The simple answer is that we run our simulations on such a long timescale in order to build reliable statistics for the possible accretion states of the black hole. In this way, following the VLBI observations and reconstruction of an image of the black hole, we can compare this image to an extensive 'library' of synthetic images of an accreting supermassive black hole. For this reason, and to use a previous analogy, we do not take a single photo of our mountain but take several photos every day for 10 years. The hope is that if we ever find ourselves with a blurred photo that could be of a mountain, our rich catalogue of images of mountains under various conditions will allow us to understand whether it is a mountain and possibly deduce its properties.

Following this logic, scientists of the EHTC, and in particular the research group in Frankfurt, have been engaged in the most extensive and systematic campaign of numerical simulations of accreting black holes ever attempted before. The aim was to construct the broadest and most complete description of the physical conditions encountered by a plasma when accreting onto a black hole. Using complex and sophisticated numerical codes and exploiting a multitude of parallel supercomputers, a large number of simulations at very high angular resolution were performed, covering all of the most important scenarios in which a supermassive black hole can accrete matter.

In particular, we carried out simulations by varying one or more of the following parameters:

- *The spin of the black hole*: considering both non-rotating and very rapidly rotating black holes.
- *The direction of rotation of the plasma in respect to the black hole*: studying plasma that is either co-rotating or counter-rotating with the black hole.
- *The strength of the magnetic field*: exploring two different types of accretion discs that differ in the accretion rate they can produce and which can be either small or large.

You may have noticed I have not mentioned the mass of the black hole as a parameter that could vary; this is because all of the black holes considered in the numerical simulations had exactly the same mass. This was possible because, although the mass of the black hole plays a very important role, it can easily be factored out of the problem. All of the quantities computed in the simulations, in fact, can be properly rescaled, so the same simulation can be used to describe the accretion onto a black hole of one solar mass or one billion solar masses.

It is not helpful at this point to discuss the many simulations performed, each of which has allowed us to produce hundreds of distinct synthetic images. Suffice it to say that the simulations covered a large variety of scenarios, each different from the others. In some simulations, the accretion disc is very bright while, in others, it has a very weak emission. In some cases, the disc terminates very near the horizon (generally, this happens for highly spinning black holes with the plasma in co-rotation). In others, it terminates very far away (this happens instead for black holes that are highly spinning but with the plasma in counter-rotation).

Finally, to further enlarge the range of possible synthetic images, each simulation was accompanied by a variety of emission models that specified the energetic properties of the electrons responsible for the radiation in the radio band. As a result, using the same numerical simulation produced a wide variety of additional realisations of the synthetic images, generating a new sequence for each of the emission models employed.

In its effort to build a theoretical interpretation of the observations and extract as much information as possible, it comes as no surprise that, in just over six months, the EHTC has produced as many as *60,000* synthetic images. These images cover the majority of possible configurations in which an accreting black hole could appear. This huge library of synthetic images is even more impressive when you bear in mind that we didn't know how M87* was oriented concerning our line of sight. Indeed, this is one piece of information we wanted to learn from observations. Hence, each synthetic image alone can generate hundreds of other images simply by changing the two angles under which the image can be rotated.

From a conceptual point of view, each of these 60,000 images represents a plausible and physically consistent realisation of how an accreting supermassive black hole might look when viewed with a network of VLBI-connected radio-telescopes. However, it is also possible that none of these 60,000 *synthetic*

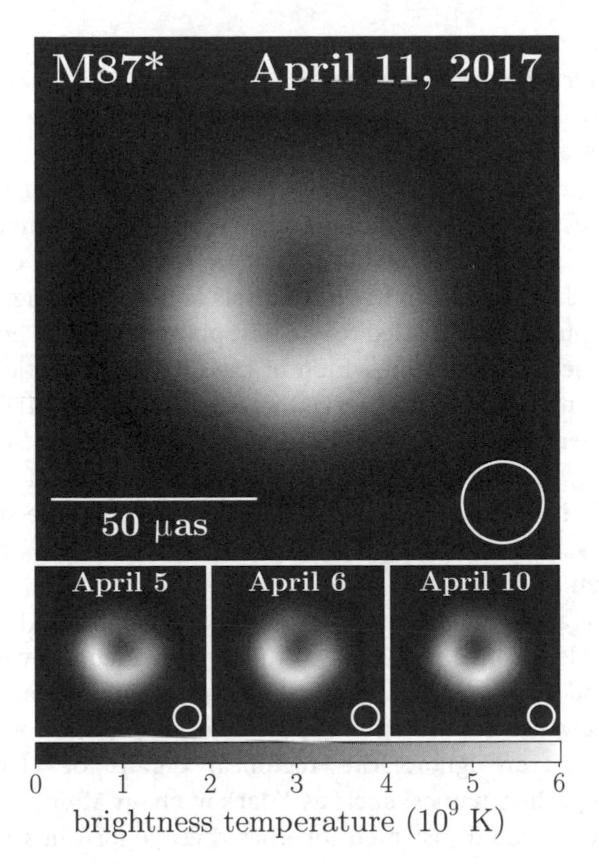

Fig. 7.12 Four different images of M87* taken on different days in April 2017 by the EHTC at a radio frequency of 230 GHz. © The EHTC.

images corresponds to any one of the four *observed* images reconstructed by the EHTC based on data collected on April 5, 6, 10 and 11, 2017, and on which I report below.

Clearly shown in Figure 7.12, and as expected, these observed images show some variation from one day to another but also reveal an overall 'robustness' of the basic properties of the images.

Like a Needle in a Haystack

Allow me to give you a recap of the long sequence of events and arguments made so far. In April 2017, after months of preparations, the EHTC performed VLBI observations of M87* and, after more than one year of work on the data, it was able to reconstruct four observed images. Meanwhile, between the end of 2018 and the beginning of 2019, the EHTC ran a massive campaign of theoretical simulations, trying to predict how a supermassive black hole surrounded by accreting plasma should look, thus producing 60,000 synthetic images. I have not yet described the final step in our analysis and the missing piece in the big puzzle that has led to the first photo of a black hole. This step is necessary to find which of the 60,000 synthetic images correspond to the four observed images so that the EHTC could finally learn what the observations were all about. Although this may seem very simple to take from a logical point of view, it is much like looking for the notorious needle in a haystack.... Fortunately, we could use computers to overcome this final challenge.

There are sophisticated mathematical algorithms that make it possible to solve problems of this type (which are very complex and characterised by multiple possible parameters) by exploiting the computational power of large supercomputers. Here, we can ignore the technical details of algorithms with appealing names, such as 'Markov chain Monte Carlo' or 'genetic'. I would just mention that these algorithms can find an optimal (or quasi-optimal) solution to problems using large amounts of data that depend on many parameters. Once applied to the data and a 'best fit' is found for the data and the model, these algorithms quantify the probability that a given parameter is contained in a certain interval. In this way, it was possible to compare the four observed images with the synthetic ones and determine the probabilities that a certain parameter, for example, the mass of the black hole or its spin, would fall within a certain interval.

To better understand how this matching process works in practice, let's perform another thought experiment. Just imagine you are in a stadium hosting 60,000 spectators who are anxiously waiting for an important game to start. On entering the stadium, you catch a glimpse of someone who looks like a good friend of yours, whom you lost in the bustling crowd at the entrance. Hence, you are not sure the person you have seen is your friend and, if it is, whether or not they are now in the stadium. Luckily, you happen to have a recent but very blurred photo of your friend with you. Security at the stadium is based on closed-circuit television cameras that capture pictures of each person as they enter the stadium. So, you approach the kind manager of the surveillance system and ask him to compare your blurred photo with the high-resolution mugshots taken at the entrance by the fixed cameras. The comparison process that the surveillance system performs will use Markov chain Monte Carlo or genetic algorithms to find the best match for the images taken at the entrance and that of your friend. In practice, facial-recognition techniques will be used, which decompose a human face into main features (for example, the distance between the eyes, the position of the nose in respect to the eyes, the size of the mouth, etc.). Essentially, each mugshot is decomposed into a series of parameters (basically N numbers), with one parameter for each facial aspect. The algorithms then compare the sequence of N numbers corresponding to your blurred photo with the N numbers for each of the 60,000 images in the library of mugshots. The 'intelligent' part of the algorithm lies in looking for the best match without accessing the entire library, but instead 'moving' within the space of N sequences in an efficient manner, concentrating only on those sequences for which there is already a good match. In this way, the search is simpler and, above all, faster.

Figure 7.13 offers a schematic representation of this process. The top left image is the blurred photo of your friend (which, at least from a logical point of view, is similar to the right-hand side of Figure 7.1), and the image in the top right is a comparison from the library of mugshots (consisting of 54,200 images to be precise).[10] The two rows below show 10 photos from the library selected by the algorithms as those closest to the reference image. From a mathematical viewpoint, the images in the rows below are those that received the highest 'scores' in the comparison of the parameters associated with your friend's face.

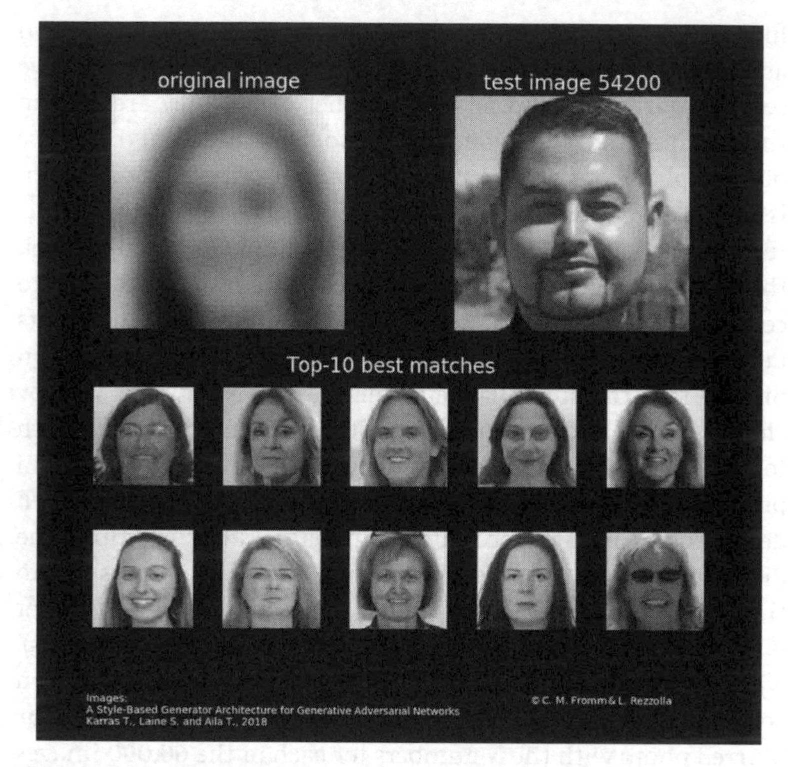

Fig. 7.13 Image comparison process. © C. Fromm/LR. Images: T. Karras, S. Laine, T. Aila.

The process of image comparison shown in Figure 7.13 (conceptually very similar to what was done for M87*) teaches us some important lessons. First, it reveals that the mathematical operation of projection of the reference image into the space represented by the library of images of spectators in the stadium is not unique; that is, the projection is degenerate. In terms that are perhaps easier to understand, the comparison process reveals that, in general, there are multiple candidates. In our case, there are at least 10 out of the tens of thousands of test images (there could be more or less depending on the tolerance of our algorithm) who, essentially with the same score, meet all of the criteria we imposed to find an image that is 'sufficiently similar' to that of your friend. No surprise here; as in the case of the wall clock, as long as the amount of information available to us is limited and restricted to observations, a multiplicity of possible answers is almost inevitable.

Second, while there are good matches, it appears that your friend is not at the stadium since none of the best images seems to correspond to the reference image, even though some are quite close. Please note, I wrote 'seems' because these face-recognition algorithms only measure the probability of a certain match, and this probability is never 100% unless we are dealing with 'perfect' images, that is, without noise or, if you like, perfectly sharp and at very high resolution. Also, in the comparison between the observed images of M87* and the synthetic ones, the result of the match is measured in terms of probabilities. Notwithstanding that, we have done our best to create a synthetic image as close as possible to reality; the latter is always more complex than we can represent with our simulations. Therefore, it is perfectly plausible and scientifically acceptable that none of the hundreds of thousands of images produced synthetically corresponds 'exactly' to the one observed.

The properties of the 10 best-matched images in the lower rows provide the third and last lesson. The selection given by the algorithms reveals that your friend belongs to a particular class of human beings: a female with long hair. This is because all of the subjects in the best-matched images are women. Although there were also men in the stadium with long hair, the face-recognition algorithm didn't pick any of them out, revealing that your friend has long hair and also is a woman.

Of course, you might argue that this result is trivial: you already know that your friend is a woman and that she has long hair. However, regarding the image of M87*, we did not know what class of object the image referred to. Hence, just by studying the properties of the best matches, we could learn a lot about the image, even if we did not find a perfect match!

Let's return now to what we have actually done with the EHTC and M87* and set aside friends and events in the stadium. Figure 7.14 illustrates this by showing the excellent comparison

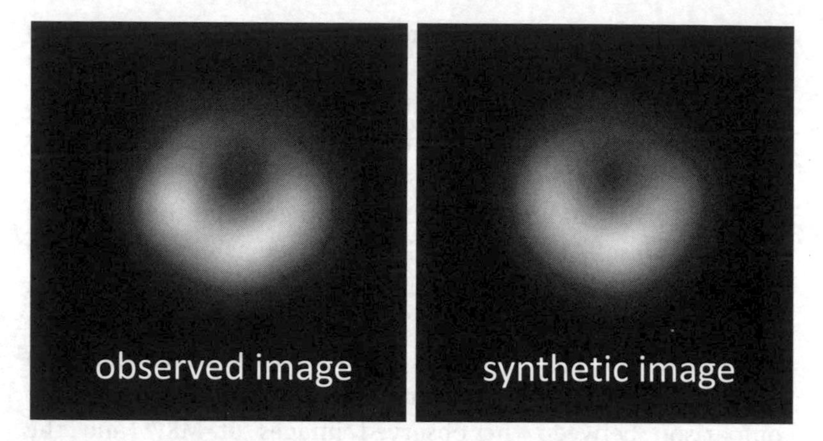

Fig. 7.14 Example of the excellent match between an observed image of M87* and a suitably deteriorated synthetic image. © The EHTC/C. Fromm/LR. A black and white version of this figure will appear in some formats. For the colour version, please refer to the plate section.

between the image of M87* observed on April 11, 2017 (left panel) and a synthetic image appropriately blurred to take into account the experimental noise (right panel).

I think it is clear that the similarity between the two is impressive. In fact, the score that the algorithms have assigned to the image on the right is among the highest found among the hundreds of thousands of synthetic images produced.

The image on the right is a fully theoretical prediction. It was produced through simulations of plasma dynamics and by employing a precise model for the emission of the radiation. Hence, we know every feature of the object behind the image on the right: the mass and spin of the black hole, the inclination with respect to the observer, the temperature and energy distribution of the electrons, etc. Logically, we could then believe that we also know everything about the image on the left. However, this logic, albeit simple, is unfortunately flawed! The simple reason is that, although the image on the right offers a very close theoretical representation to the observed reality, it is unfortunately not unique. Just as for the women shown in the 10 small panels in Figure 7.13, in fact, there are also degeneracies in the case of M87* that cannot be removed and that are the direct consequence of the finite (basically low) resolution at which the observed image has been constructed.

The most important consequence of this degeneracy is that we have found a number of synthetic images that match equally well-observed images but refer to very different models of black holes, for instance, to a non-rotating black hole or to a black hole that is maximally rotating. This result, which at first glance might seem impossible, is instead perfectly plausible; it is the consequence of using different models for the emission of radiation. It is thus possible that two different emission models applied to two different accretion models conjure up almost identical low-resolution images. In other words, the differences in spin and emission can combine in such a way as to produce two images that our algorithms recognise as being very similar.

This result is not too surprising given the novelty and quality of the first observations of M87*, and it is both 'bad' and 'good'. It is bad in the sense that it reveals that, with our observations, we are unable to appreciate the differences introduced by the spin of the black hole and its orientation (in other words, our observations are 'insensitive to spin'). Consequently, we cannot really measure the spin of M87* and, indeed, the EHTC publications in 2019 do not report any estimate for the spin of M87*. At the same time, this degeneracy is good because it lets us conclude with great confidence that the right-hand image in Figure 7.1 does indeed show a supermassive black hole. In fact, all of the synthetic images that best match the observations always correspond to a simulation of an accreting black hole and, under all conditions, yield an image as in the right-hand side of Figure 7.1, regardless of the value of the spin.

This is exactly what happened in our thought experiment at the stadium; the face-recognition algorithms didn't inform us with any certainty about the presence of your friend. However, they did inform us, with very high confidence, that your friend belongs to the class of human beings that are women with long hair and not men with long hair. Similarly, and notwithstanding the degeneracy between theory and observations, we can conclude that the image of M87* most likely belongs to the class of compact objects that are black holes, for which it has all the basic features. And this is indeed a revolutionary result!

I want to conclude this section with two optimistic considerations. The first is that I didn't really tell you the whole story when I said we are wandering in the dark concerning the spin of M87*. Although we are insensitive to spin if we limit ourselves uniquely to the EHTC images, we have other cards up our sleeves. We can also consider the observations made in the past by other scientists about the central object in the M87 galaxy and this would immediately shed some light on its spin. In particular, we could use the power emitted by the relativistic jet of M87*, which can be measured with good precision given the huge dimensions of the jet. In this way, we can rule out that M87* is a non-rotating black hole. This is because all of our numerical simulations show that Schwarzschild-type black holes do produce jets, but these are energetically 'too weak' to explain the jet power measured by the observations. In essence, with the help of the observations made by other astronomers, we can conclude with reasonable certainty that M87* must be a rotating black hole, even if we do not yet know its rotation rate or its direction of rotation.

The second optimistic consideration is that, as explained in many of the examples I have given, the degeneracy between theory and observation is not destined to last forever. The images obtained are simply the *first* of M87*, and we will be able to produce more images with ever-increasing precision in the future. Unlike the observations of black holes made through gravitational waves (which will be discussed in the next chapter) that can be carried out only once, we will continue for many decades to take photos of the remarkable black hole, M87*. These new observations will be more and more accurate and detailed, not only because they will be made at ever shorter radio wavelengths (let's not forget what Equation (7.1) taught us) but also because they will be made by an increasingly large number of radio-telescopes. Finally, when imagining a future that is perhaps not too distant, these radio-telescopes will be mounted on satellites and then connected with the VLBI technique, thus creating baselines of tens of thousands of kilometres between them, that is, many times the diameter of the Earth.

In short, the inevitable veil of uncertainty that, for now, surrounds the value of the spin of M87* will not remain there for long; we will soon find out!

Are We Really Sure?

We have seen that the observations made by the EHTC are in agreement with the predictions of the theory, based on models of accretion onto and emission from black holes. It follows that the very concept of black holes as expressed by the theory of general relativity is perfectly compatible with the observations.

Nevertheless, on several occasions, we have paused to reflect that astrophysics, as an observational science, must live with the inevitable possibility of a degeneracy between the observations and the models capable of explaining them. We have also seen that the black-hole solution is not the only one, even though it represents the most natural and simple explanation to the phenomenology observed in compact astrophysical objects. Alternative interpretations involve very compact objects with or without a surface and not having an event horizon. Finally, we have seen how Einstein's theory of general relativity, whose predictions have always proved correct, does not represent the only theory of gravity capable of explaining the observations.

Based on these considerations, you could take one of two stances. A more conservative stance would assume the only correct interpretation of the EHTC observations to be the simplest and most natural and would involve the concept of black holes in general relativity. In this case, there would be no need for you to read any further, and you could immediately skip to the next chapter. The second position, more open and less conservative, is to be aware of an important lesson that history has taught us so far: what appears exotic and abstruse today can become plausible and acceptable tomorrow. If you feel more inclined towards the second attitude, then you should continue reading because I will try and address a question that is as simple as it is fundamental: 'Are we sure that what was imaged by the EHTC is a black hole?'.

Personally, and notwithstanding that I have devoted my entire career to studying the consequences of general relativity (and I firmly believe that this theory, however incomplete, offers the correct interpretation of gravity), I am definitely among those with a more open-minded attitude. This is why, together with my collaborators, I conducted a series of studies aimed at answering the question above. Obviously, I cannot go into the details of such research here, the technical and mathematical aspects of which are a poor fit for a popular-science book. However, I can share the most important conclusions, and I will do this in the remainder of this chapter.

In essence, the name of the game that my collaborators and I have decided to play is to extend to other compact objects, either in Einstein's theory or other theories, the analysis carried out so far on accreting black holes in Einstein's theory. And hence, to determine to what extent it is possible to distinguish such objects based on the observations. For this purpose, we carried out the same plasma dynamics simulations we discussed in the cases of Schwarzschild and Kerr black holes. We also used the same models of radiation emission with the final goal of obtaining synthetic images of these 'exotic' compact objects. Finally, we used the same information about the properties of radio-telescopes and their resolution to obtain realistic radio images.

In our comparisons so far, we have considered two classes of compact objects. The first is a *dilaton black hole* and can be seen as a generic representative of a class of compact objects that can be an alternative to Einstein's black holes, which also has an event horizon. On the other hand, a compact object of the second class is represented by a so-called *boson star*, which does not have an event horizon, but has a compactness comparable to black holes. For those who may be wondering, a dilaton black hole emerges as a solution in the Einstein–Maxwell dilaton–axion theory of gravity. This theory couples the Einstein–Maxwell equations to the dilaton scalar field and the axion vector field. In essence, the presence of these two additional fields breaks the assumption of a vacuum being needed to obtain a black-hole solution in general relativity and provides the black hole with additional properties ('hairs'). The existence of these two fields is entirely speculative at this point. A boson star is instead a self-gravitating cloud of a scalar field (again of underdetermined nature and origin) with a core that can be very compact, effectively appearing as a black hole in terms of gravitational effects. Our choice of a dilaton black hole as a representative case for an alternative theory was motivated by such a solution being physically very distinct from general relativity, as it requires the presence of an additional scalar field (i.e., the dilaton field), absent in Einstein's theory. Similarly, the choice of a boson star is linked to this type of object being discussed often in astrophysics as a possible candidate for dark matter, and because it should have a behaviour very similar to that of a wormhole, another popular exotic compact object we have encountered.

Skipping all the details related to these studies, which you can be sure are numerous and can be found in the corresponding papers,[11] The final comprehensive result of this research is given in Figure 7.15. In particular, the figure shows synthetic images, at a frequency of 230 GHz, of the three compact objects when they are set to have the same mass and are subject to the same

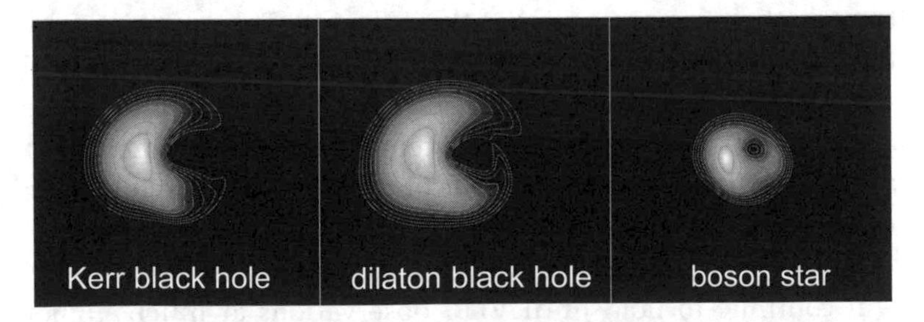

Fig. 7.15 Comparison of the images at 230 GHz of three compact objects with the same mass and subject to the same accretion processes. From left to right: a Kerr black hole, a dilaton black hole and a boson star. © C. Fromm/Y. Mizuno/H. Olivares/Z. Younsi/LR. A black and white version of this figure will appear in some formats. For the colour version, please refer to the plate section.

accretion processes via a geometrically thick disc. The left panel refers to a rotating black hole in Einstein's theory (i.e., a Kerr black hole with spin $a = 0.7$). In contrast, the centre panel shows a dilaton black hole with a dilaton parameter 'b', such that it has an innermost stable circular orbit equal to that of the Kerr black hole (i.e., $b = 0.5$). Finally, the panel on the right refers to a boson star having a compactness of $M/R \simeq 0.1$, close enough to that of a black hole.

The beauty of Figure 7.15 is that it doesn't need much explanation. It is quite clear that, while it is possible to notice differences between either of the black holes (left or central panel) and the boson star (right panel), this is not true when comparing the two black holes. In essence, our research has led us to conclude that it is currently possible to distinguish a black hole from a boson star. A boson star shows a much smaller shadow than a black hole of the same mass (a similar behaviour is to be expected for a wormhole, i.e., a smaller shadow for the same mass). Furthermore, given the fairly tight constraints on the mass of M87*, which do not allow the size of the measured shadow to vary too much, a boson star can be excluded, at least in the model considered so far, from being compatible with the EHTC image. At the same time, given the current quality of the image of M87*, it is not possible to distinguish with sufficient certainty a Kerr black hole from a dilaton black hole, even though their images are slightly different. This explains why this chapter has the title 'The First Image of a Black Hole' and not 'The First Image of an Einsteinian Black Hole'. At present, we are not able to distinguish an Einsteinian black hole from one in a dilatonic theory and probably also from those predicted in many other theories.

It is possible, and perhaps even likely, that these conclusions will continue to hold until VLBI observations at much higher resolution are carried out. However, it is also very likely that new and repeated observations, as are expected in the coming years, will overcome some of the degeneracies in Figure 7.14. The simulations clearly show that the variability in the accretion rate is different in different black holes, so comparing images taken at varying times can be decisive in breaking the degeneracy now present when comparing only two static images. Once again, the images taken so far of M87* should be considered simply as the first ones; numerous questions remain, which may be answered as new measurements are carried out.

I want to end this chapter, dedicated to the first image of a black hole, with a general consideration. We have already mentioned how the EHTC observations have marked a technological revolution and enormously enriched our understanding of astrophysics and gravitational physics. In particular, at least two scientific results, generated by the EHTC observations and their theoretical interpretation, are unanimously recognised as the most relevant. The first is that they have provided evidence, beyond any reasonable doubt, that there is an accreting supermassive black hole at the centre of the M87 galaxy. Since M87 is not a galaxy with peculiar characteristics, by induction, these observations tell us that every galaxy should have a supermassive black hole at its centre. It is a hypothesis as old as the first observations of galaxies and black holes and, remarkably, only confirmed in 2019. Even more important, perhaps, is that the confirmation of this hypothesis is essential for our current understanding of the formation of large-scale structures of the universe.

The second universally recognised contribution of the EHTC observations is that they provide evidence of an essential property of the black-hole solution in general relativity: its simplicity! It is indeed extraordinary that an object like M87*, with a mass of a few billion solar masses, has exactly the same properties as an object whose mass is a hundred million times smaller, such as the black holes revealed by the LIGO and Virgo gravitational-wave detectors. The simple scalability with mass of black-hole behaviour that we have encountered with the no-hair theorem is a precise prediction of general relativity. The EHTC and LIGO–Virgo observations are so beautifully complementary; the reason is that they are both necessary to explore the concept of black holes at the largest and the smallest scales at which they are expected to exist.

I would add one final consideration that reflects a very personal view: the most profound relevance of the EHTC observations has to be sought elsewhere. What I find most valuable about the first image of a black hole is that it has transformed the event horizon from a 'mathematical concept' into a 'physical object'. In other words, the observations have shown that it is now possible to carry out experimental tests on what was previously only a concept and, as such, subject to a variety of different views. Paraphrasing Galileo, these observations, and all those that follow, essentially allow us to stop '[arguing] about the greatest issues without obtaining any truth'. It is now possible to use actual observations to compare different and equally plausible theories and thus accept those compatible with the observations and reject those that are not. The possibility of carrying out this trivial but essential step in the scientific method has been denied us so far. Therefore, the EHTC observations have ultimately provided the simplest and most important tool for a scientist: the possibility of arriving at truths. In this, I recognise the greatest value of the M87* observations.

8 GRAVITATIONAL WAVES: CURVATURE IN MOTION

Since their direct detection in 2015 via the merger of a binary system of black holes, and almost a hundred years after Einstein predicted them to be a weak-field solution of his equations, gravitational waves have now become 'common knowledge'. While most of you may have heard of them, I wouldn't be surprised if they remain as something obscure and leave you wondering: 'Yes, but *what are* gravitational waves?'. Now that we have a much better understanding of gravity, we can answer this question properly.

Let's start with the correct and concise answer: gravitational waves are solutions of the *linearised* Einstein equations, that is, valid in situations when the spacetime curvature is small. They are also *wave*-like solutions and, consequently, represent the propagation, at the speed of light, of 'ripples' in the spacetime curvature. I am aware that, albeit correct, this answer may still be obscure for most. Therefore, in the course of this chapter, I intend to explain what the answer actually means in what I hope is a more intelligible manner.

To do this, I will necessarily have to start from the notion of a wave, which is a fundamental concept in physics. From a mathematical point of view, we speak of a wave whenever we have to deal with specific types of partial differential equations, which take the name of wave-like equations. These equations are so widespread and frequent that we have even adopted a special mathematical symbol to represent them, namely the 'box' symbol $\Box$.[1] From a physical viewpoint, a wave is a phenomenon in which something is transported in space and time from one point to another. That is to say, a phenomenon in which a certain physical quantity moves following precise rules as it propagates in spacetime. There are many possible examples of wave phenomena: sound waves, seismic waves, electromagnetic waves, water waves, etc. In all these cases, slight variations of an otherwise constant physical quantity propagate from one position in space to another. For example, let's consider sound waves, which allow us to satisfy some of our primary needs, such as communicating or listening to music. Well, they are nothing but small perturbations in pressure and density that propagate in a fluid (gaseous or liquid) at a precise speed, namely, the 'speed of sound'. Density perturbations also propagate with seismic waves, such as those regrettably produced by earthquakes or those employed, at much smaller amplitude, to probe the properties of terrain both horizontally and vertically.

Similarly, we are dealing with waves every time we talk of photons, as these are perturbations of electric and magnetic fields that propagate at the speed of light. Going back to the example of water waves, these are produced when a stone is thrown into a pond. In this case, too, we have the propagation of small perturbations, which are slightly more complex as they develop on the separation surface between two fluids with different properties: water (an incompressible fluid) and air (which is compressible).

In all these examples, waves are understood to be small variations of a physical quantity in respect to its reference value that propagate away from their source at a precise speed, which is different for different types of waves. In addition, all these waves carry with them both energy and momentum, producing changes that can be perpendicular to the direction of propagation (*transverse waves*, such as electromagnetic waves) or along the direction of propagation (*longitudinal waves*, like sound waves). Finally, some waves can be a combination of both transverse and longitudinal waves, such as water waves.

Although all wave phenomena are similar, since they satisfy the same type of equations, gravitational waves deserve separate discussion, as they involve the perturbation of something decidedly unusual in our experience.

Waves Like the Others, but Different

The basic concept of a wave should be clear enough, partly because it is a phenomenon we commonly experience, but that of a gravitational wave is undoubtedly less clear, as it involves something we are not used to, the curvature of spacetime. It may be a good idea to take a step back to understand what this is all about and return to the sheet and bowling ball analogy presented in Chapter 3. In particular, I would like to draw your attention to an important detail of that example: it refers to a *static* scenario. The curvature appears at a certain point because a certain quantity of matter (or energy) is present that does not change. However, it should be evident that if we moved the source of the curvature (in our case, the bowling ball) from one point of the sheet to another, the deformation of the sheet would follow this movement.

This apparently trivial detail is actually very significant, as it reveals that curvature should not be considered as a *static* property of spacetime but, on the contrary, as a *dynamic* feature and, as such, it can change position, that is to say, 'move', in spacetime.

And if it seems obvious that the curvature should move when the source that generates it moves, it should be equally plain that it can change and be transmitted when, for whatever reason, the properties of the source change. In other words, following this line of logic, we would expect that a source of curvature, which changes its properties because, for example, it moves from one point to another, or because it varies in shape or compactness, would also generate perturbations of curvature that propagate in the form of waves.

Let's take the example of the sheet. Gradually pushing the bowling ball from one point to another on the sheet's surface would produce small folds in the fabric. These folds would continuously change as we move the ball and, however tiny, would reach the edge of the sheet. Well, gravitational waves are just like that: curvature perturbations that propagate when a source is no longer static. In other words, and as the title of this chapter suggests, gravitational waves should be thought of as *curvature in motion*.

Someone with a particularly critical disposition could argue that something funny is going on here. We have learned that any object with mass (or energy), and therefore even a grain of sand, is capable of producing curvature. It follows that it would be sufficient to take a handful of sand, let it flow through our fingers and create gravitational waves. Admittedly, this is difficult to believe, yet it is exactly what happens! A falling grain of sand, the flight of a butterfly or even our gesticulations as we speak represent sources of gravitational waves. The theory of general relativity leaves no room for doubt in this respect: gravitational waves are solutions of Einstein equations in weak gravitational fields (that is, sheets that are almost flat) which, once created by a source, propagate at the speed of light, generating transverse perturbations. To reconcile the fact that gravitational waves are easy to produce with the lack of evidence for their existence in our daily experience, it is necessary to remember that the strength of the curvature is key. We have seen how spacetime is fundamentally rigid and resists being curved, except by compact objects, such as neutron stars or black holes. A grain of sand or a butterfly do produce curvatures, but these are tiny. Likewise, the gravitational waves generated by the motion of a grain of sand or the flight of a butterfly will be so imperceptible that I doubt there will ever be a device capable of measuring them. So, there really is nothing funny going on: gravitational waves are easily produced by dynamic changes in the curvature. However, objects with different mass–energy will produce waves whose amplitude is proportionate to their mass–energy.

Now that we understand a bit better what gravitational waves are and how easy it is to produce them, we could ask ourselves: 'Given a source of curvature, what type of changes can produce gravitational waves?'. The answer is straightforward: *all of them!* In other words, any possible variation in the properties of a source of non-negligible curvature will generate gravitational waves. The one exception to this simple rule is deformations that maintain perfect spherical symmetry. This result, linked to the Birkhoff theorem introduced in Chapter 6, is particularly interesting. It tells us that if we take a Schwarzschild black hole with a

Fig. 8.1 Numerical simulation of gravitational waves emitted by a perturbed black hole. © R. Kaehler/LR. A black and white version of this figure will appear in some formats. For the colour version, please refer to the plate section.

spherical symmetry and generate a spherical-type perturbation by, for example, letting matter fall on it equally only along radial directions (i.e., via a spherical accretion), it would not emit gravitational waves. Conversely, if we took the very same black hole and exposed it to a non-spherical accretion (for instance, via an accretion disc), it would emit gravitational waves.

Waves of this type are shown in Figure 8.1 and are the result of a numerical simulation. The image details the gravitational waves produced in the first moments in which an almost spherical neutron star collapses into a black hole that is not yet perfectly spherically symmetric but has small perturbations shedding away.

Putting aside the case of spherical symmetry (which is an exception as it is quite rare in nature to maintain a perfect spherical symmetry), it must be said that, although gravitational waves are easy to generate, not all sources can produce them efficiently enough or at frequencies that allow for experimental detection. We return to these two aspects, the efficiency of emission and detection capability, later on in this chapter. For the time being, let's concentrate instead on their effect on the rest of spacetime when they propagate.

To explain what happens when we interact with a gravitational wave, I will employ an analogy with a type of wave closer to our lived experience: the water waves generated on the perturbed surface of a lake. So, let's imagine we are on an inflatable mattress, floating peacefully on the surface of a perfectly calm lake on a sunny and windless day. At one point, we see someone jump from a boat in the distance. Their impact with the water produces a perturbation that spreads outwards, with an essentially spherical front and a decreasing amplitude, as the wavefront is distributed over an increasingly wider surface. Once it reaches us, the perturbation is very small, and its wavelength (the separation between successive crests in the wave) will be much greater than the size of our inflatable mattress, which we can, therefore, consider as point-like (i.e., with vanishingly small size). As a first approximation (or, in more mathematically precise language, at first order in the perturbation), we can say that the waves will be transverse, so our mattress will begin to oscillate with a periodic motion in the vertical direction, i.e., perpendicular to the direction of propagation of the wave. In this example, the water waves essentially do the job of transferring the kinetic energy released by the person who jumped from the boat to the rest of the lake. Once the wave-train (that is, the sequence of waves triggered by the jump) has passed completely, the mattress will return to its initial position, at rest, and we can go back to relaxing.[2]

By analogy with the above description, we can perform a thought experiment. Imagine that, somewhere in the universe, gravitational waves are produced (through processes described further on in the chapter), which reach us along a direction that is perpendicular to the lake's surface, with a wavelength that is again larger (though not by much) than the size of our mattress. Since gravitational waves transverse, the perturbation they propagate cannot be measured as a variation in the vertical position of the mattress. Rather, they generate local changes in the value of the curvature. More precisely, by changing the local curvature, gravitational waves will generate 'tidal forces', which will then induce variations in the position of objects not subjected to external forces. This is similar to the tidal forces exerted by the Moon on the Earth, which cause the displacement of liquid masses on our planet. Thus, the tidal fields that would reach us in our thought experiment would induce deformations both on us and, above all, the water around us. These deformations are of a very specific type, namely, *quadrupolar*, and consist of a 'compression' in a given direction and a 'stretching' in the orthogonal direction. As we will discuss later, these variations will be periodic and emitted at the same frequency as the waves, which can either be constant or vary over time.

To better understand what we would be experiencing in our thought experiment, we can rely on Figure 8.2, which shows the quadrupolar deformations to which the *Vitruvian Man* by Leonardo da Vinci would be subject if crossed by a gravitational wave. Fortunately for us, the gravitational waves that reach the Earth are so weak that not only do they not produce appreciable deformations in our bodies, but it is actually necessary to build highly sensitive devices to detect them.

It is important to comment here on an aspect that is not often given proper relevance. In their mathematical formulation, the Einstein equations only behave like waves under precise conditions: a vacuum spacetime, a weak field and asymptotic flatness. These conditions are certainly not those present in the vicinity of the source, where spacetime has a strong and very dynamic curvature. However, these conditions are met from where we are, namely, at a great distance from the sources and where these perturbations do behave like waves. For us, as distant observers, curvature perturbations will therefore behave like waves and come in two types, or two degrees, of *linear polarisation*. These are called *plus polarisation* (indicated with

Fig. 8.2 Quadrupolar deformation of the *Vitruvian Man* crossed by a gravitational wave. © M. Pössel (AEI)/LR. A black and white version of this figure will appear in some formats. For the colour version, please refer to the plate section.

a + symbol) and, identical to the plus polarisation but rotated by 45 degrees, *cross polarisation* (indicated with a × symbol). To get a concrete idea of the difference between the two polarisations, we can refer again to Figure 8.2, which shows the effects of plus polarisation. By contrast, the cross polarisation would not compress and stretch the *Vitruvian Man* in the direction between his head and feet but, instead, would compress and stretch him along a diagonal direction to this.[3]

In general, the degree of polarisation of gravitational waves reflects the properties of the source. Realistic astrophysical sources will produce radiation containing a mixture of these polarisations, although it is possible to generate gravitational radiation in a single polarisation under some special conditions. Given these properties, let's now focus on why it is important to study gravitational waves.

Gravitational and Electromagnetic Waves

It is often remarked that the combined detection of electromagnetic and gravitational waves has marked the birth of a new multi-messenger approach in astronomy, thus opening a new 'window' on the universe. Both of these statements are correct, although a little obscure, perhaps.

Let's see what they signify and start by clarifying what 'multi-messenger' astronomy means. The formal definition is that it is a branch of astronomy in which scientific knowledge is reached through different channels of information, called 'messengers'. More specifically, in addition to the obvious electromagnetic channel (i.e., the light emitted by celestial objects in the various bands), multi-messenger astronomy makes use of other messengers, such as neutrinos (for instance, those emitted in supernova explosions) or gravitational waves. In the latter case, it also takes the name of *multi-messenger gravitational-wave astronomy*. The importance of multi-messenger astronomy is that, by using more than one channel to obtain information and because the information contained in the various channels can differ substantially, it has the potential of being much 'richer' than traditional astronomy.

To better illustrate what I mean, let me offer an example that I find fits very well and is easy to understand. Imagine taking part in one of the many beautiful festivals that characterise the Italian summer, and that the programme includes a late-night fireworks display. Because you are passionate about fireworks, you have carefully chosen the best spot to take in the sights. A position that offers an excellent and elevated view, far enough away to observe the whole show but near enough to enjoy the full experience of the blasting sounds. In essence, imagine sipping a glass of wine immersed in the warmth of the summer night, enjoying an unforgettable fireworks display.

Now, think about what this experience would be like if you were deprived of either sight or hearing. If you could not discern the lights produced by the fireworks, you would perceive the entire event as a sequence of explosions, some intense and low, others weaker and more high-pitched, which would convey little of how spectacular the fireworks were. If, on the other hand, you could not hear the sound produced by the fireworks, you would find yourself watching a sequence of colourful and perhaps fascinating, but not very engaging, images as if everything was taking place on a different planet. On several occasions, I have happened to observe fireworks from a low-flying plane, and I can assure you that it is much less exciting.

This example illustrates how fireworks can be considered, for all practical purposes, as a multi-messenger phenomenon. This is because the information from fireworks reaches us via both sound waves (the first messenger) and via electromagnetic waves (the second messenger). Thanks to this analogy, you can fully and directly appreciate how much richer and more informative multi-messenger astronomical observations can be when compared to traditional ones. They allow us to investigate different aspects of a source because they are conveyed by different messengers.

Concentrating on how gravitational waves differ from other messengers helps us to appreciate how gravitational waves represent a very different sort of messenger. The vast majority of the information we receive about our universe comes to us through electromagnetic waves. They cover a very broad spectrum, from the radio band to the gamma-ray band, but they have one property in common: they reveal details about the *thermodynamic* state of the sources, namely, their density and temperature. At the same time, the components of light (photons) interact with the matter they encounter on their way to us, therefore experiencing either absorption or additional emission. Indeed, an astronomical source could be partially or completely eclipsed by the material interposed between us and it. Gravitational waves are somewhat different because they transmit a tidal gravitational field, so they are only loosely coupled to the forces that hold astronomical objects together. As a result, they propagate almost undisturbed, providing us with information on the global movements of the sources at frequencies very different from those of electromagnetic waves; in this way, the information that gravitational waves convey is necessarily different and thus complementary.

Electromagnetic waves	Gravitational waves
They are oscillations of electric and magnetic fields that propagate through spacetime.	They are oscillations of the spacetime itself.
The simplest are those relative to dipole radiation, that is, generated by an oscillating dipole (two opposite electric charges).	The simplest are those relative to quadrupolar radiation, that is, generated by an oscillating mass quadrupole (a non-circular distribution of mass).
They are almost always the result of the incoherent superposition of contributions from millions of electrons, atoms or molecules.	They are produced by the coherent movement of large amounts of mass and energy, be they in the form of astronomical objects or energy condensations.
They show us the details of the thermodynamic state (temperature, density) of the sources that produce them.	They show us the details of the dynamics and global properties (size, mass and velocity) of large concentrations of mass–energy.
Their wavelengths are very short compared to the size of the astronomical sources. Hence, it is possible to obtain 'images' of the sources.	Their wavelengths are often comparable to the size of the sources that produce them, if not greater. Hence, it is not possible to produce 'images' of the sources.
They can be absorbed, diffused and dispersed by the matter they encounter on their path.	They propagate undisturbed, at the speed of light, through all types of materials.
Those emitted by celestial objects have a spectrum that is essentially independent of the mass of the source; for instance, the emissions from stellar and supermassive black holes are very similar.	Those emitted by celestial objects have a spectrum that depends on the mass of the source: the frequency of gravitational waves from stellar black holes is much higher than that from supermassive black holes.
They have wavelengths shorter than the curvature radii of the objects	They have wavelengths comparable if not greater than the curvature

(*cont.*)

Electromagnetic waves	Gravitational waves
they encounter as they propagate. For this reason, they experience gravitational lensing, that is, distortion and amplification.	radii of the objects they encounter while propagating. For this reason, gravitational lensing is less pronounced and more difficult to measure.

The table above summarises the sometimes radical differences between electromagnetic waves and gravitational waves and the type of information they convey.

After this synthetic comparison, I think it is clear that the possibility of observing the same astrophysical phenomenon through the emission of gravitational waves and electromagnetic waves represents a fantastic opportunity to multiply and amplify the information collected. The first gravitational-wave event detected was GW150914, whose name underlines that this is a gravitational-wave (GW) event detected on September 14, 2015. The signal was produced by a binary system of black holes and, for months after the detection, astronomers searched for an 'electromagnetic counterpart', that is, for traces of a corresponding electromagnetic emission in all possible bands, from gamma ray to radio. Unfortunately, no electromagnetic counterpart was identified, and this remains the case even now, several years later. Similar failures to detect an electromagnetic counterpart have accompanied other gravitational-wave detections coming from binary systems of black holes (at present, there are over 60). Fortunately, this rather frustrating situation changed in 2017 with the event GW170817: the merger of a binary system of neutron stars that gave us a fantastic display of cosmic fireworks. We will talk about this event soon.

How Do We Detect Gravitational Waves?

We have seen that the production of gravitational waves is by no means a rare or complex phenomenon, quite the contrary. What makes them so unusual in our experience is that the amplitude of the gravitational waves we usually receive on Earth is not large enough to be measured by an experimental device. To understand why, we must first remember that the propagation of a gravitational wave manifests itself through tidal fields that induce a variation in the size of the objects encountered. Shortly, we will discuss in more detail what produces gravitational waves with an amplitude that allows us to reveal them, but I can mention here that, even considering the strongest astrophysical sources imaginable (a binary system of black holes) and taking into account the typical distances of these sources, the amplitude, h, at which their gravitational waves reach us is of the order of only $h \simeq 10^{-21}$, a number so small it is equivalent to one part in 1,000,000,000,000,000,000,000. It is hard to imagine the precision required to measure such a small number, but it is as if we were to measure a length equal to one billionth of the diameter of an atom, or, equivalently, the distance between the Earth and our nearest star, Proxima Centauri, with an accuracy of 40 microns!

This precision would seem impossible to achieve in practice. Yet, after decades of research and continuous improvements, modern gravitational-wave detectors, such as LIGO or Virgo, are now able to operate at this level of precision. So, 'How do they achieve similar results?'

The answer is found in the use of a technique we discussed in the previous chapter: interferometry. In the case of gravitational-wave detectors, the basic operation involves a relatively simple device, based on an experiment conducted at the end of the nineteenth century by the American physicists Albert Michelson and Edward Morley, aimed at measuring the motion of the Earth through a hypothetical medium, the 'aether'. Unfortunately, the experiment showed that this medium does not exist. However, the very ingenious experimental device, known as a *Michelson–Morley interferometer*, remained and was later adopted for gravitational-wave detection.

Figure 8.3 shows a simplified diagram of such an interferometer. As you can see, a light source as coherent and stable as possible produces a beam of light (today, powerful

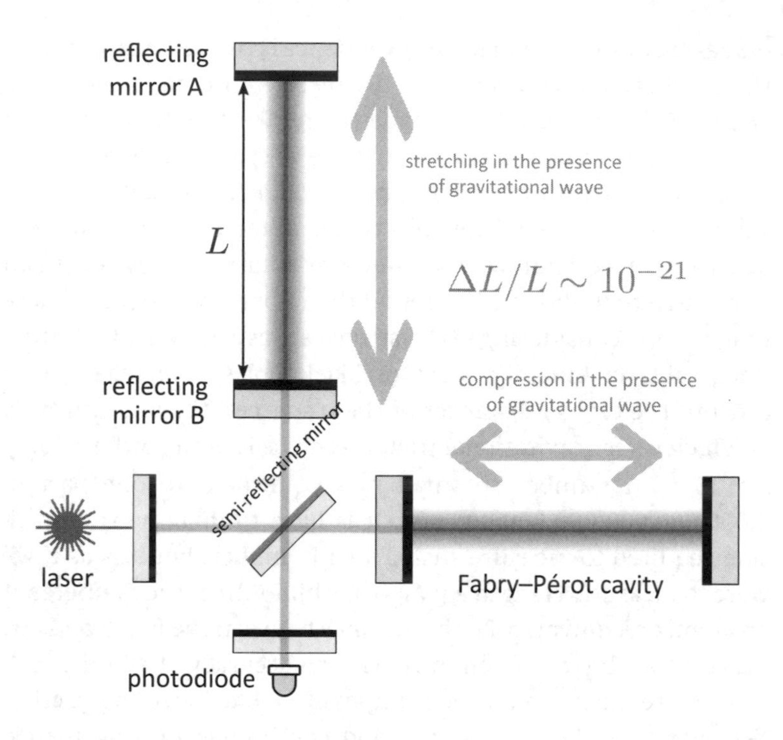

Fig. 8.3 Scheme of an interferometric gravitational-wave detector.

lasers are used but, at the end of the nineteenth century, technology was more rudimentary); this beam of light is collected by a semi-reflective mirror. Such mirrors are capable of reflecting only a part of the incident light, letting another part pass through, just like those in police stations used to observe an interview room without being seen. Regarding gravitational-wave interferometers, very sophisticated mirrors are used to reflect precisely 50 percent of the light, allowing the remaining 50 percent to pass through. Furthermore, suppose the mirror is placed at 45 degrees to the incident beam, as shown in the figure. In that case, it 'splits' the beam in two (in fact, they are also called 'beam splitters'), letting light propagate in two orthogonal directions, which represent the so-called *arms* of the interferometer (the arms can be made virtually longer by adding a new mirror and creating what is called a Fabry–Pérot cavity).

In the original experiment of 1887, the length of the two arms was of the order of about 10 metres, decidedly insufficient to measure gravitational waves. Since the relative variation of the arms induced by a gravitational wave (normally indicated as $\Delta L/L$) is equal to the amplitude of the wave itself (which we have seen to be $h \simeq 10^{-21}$), current technology only allows measurements to be made if the length of the interferometer arms, L, is of the order of a few kilometres. For this reason, the LIGO interferometric detectors (one located in Livingston, Louisiana and the other in Hanford, Washington State) have arms of four kilometres. The Virgo detector in Cascina, near Pisa, has slightly shorter arms of three kilometres.

The beam splitter divides the initial laser beam into the two orthogonal beams that travel along the arms until they reach the first reflecting mirror suspended with thin silicon wires ('mirror A' on the vertical arm), which reflects the incoming beams. As you can imagine, these mirrors are more elaborate than the ones we are used to; in fact, they are small technological marvels. In addition to being nearly 20 centimetres thick and weighing several tens of kilos, their surface is so smooth that their roughness is of the order of a nanometre, that is, a few billionths of a metre. Furthermore, if the mirrors in our house have a reflectivity of between 80 and 90 percent, theirs is greater than 99.9999 percent. In other words, only one photon in a million is absorbed rather than reflected.

Once reflected by these astonishing instruments, the laser beams return, join, pass again through the beam splitter, and interact at an interferometric level with a very light-sensitive electronic device called the *photodiode*. In general, in the absence of gravitational radiation, the interferometer can be 'tuned' so that, upon their return to the beam splitter, the two light beams are either perfectly *in phase* or *in phase opposition*. In this way, they can produce a constructive interference when they are in phase (their light adds up) or destructive interference when they are in phase opposition (their light cancels). For practical purposes, it is convenient that the interference measured by the photodiode is destructive and the whole interferometer is, in fact, tuned so that the photodiode is on the 'dark fringe'.

The passage of a gravitational wave propagating in a direction orthogonal to the plane containing the two arms produces a tidal distortion and thus a relative variation of the arm lengths. In turn, this will produce a phase shift between the paths travelled by the light in the two arms and hence a variation of the interference at the photodiode, which will no longer be synchronised on a dark fringe. The greater the amplitude of the gravitational wave, the greater the distortion of the arms, and therefore the phase shift in the light recombined at the photodiode.

I have just described, carelessly leaving out some important details, the basic functioning of a classic Michelson–Morley interferometric experiment. However, if we managed to build an interferometer of this type, with four-kilometre-long arms, we would not detect any gravitational waves! This is because the deformations they produce, and therefore the amplitude of our *signal*, would still be too small compared to those that are inevitably produced by the environment, which represent so-called *noise*. Since we cannot change the amplitude of the signal, which is entirely and solely determined by the properties of the source, our only hope of detection comes from reducing the noise as much as possible. For us to do this, it is necessary to resort to a whole series of experimental measures aimed at reducing the noise level below that of the signal, that is, increasing the signal-to-noise ratio. It is similar to being in a room and hearing a tune that you recognise but can't place because the noise made by other people in the room prevents you from clearly distinguishing the notes. Not being able to increase the music volume, your only option is to reduce the noise, so you ask the others in the room to be quiet for a moment.

Numerous techniques have been developed over the past 40 years to break down the various sources of noise that affect modern gravitational-wave detectors. As interesting and ingenious as these techniques are, discussing them would take more time than we have left on this last part of our journey. What matters here is to underline the fact that detectors such as LIGO and Virgo represent the result of the joint work of thousands of people and, as for EHTC radio-telescopes or the accelerators at CERN, they should be seen as the apical expression of human technological capabilities. Ultimately, these detectors represent the only tool available for detecting the sources of gravitational waves, which we will discuss in the next section.

Who Makes the Waves?

The title of this section is intentionally incomplete. By now, you know that any non-spherical variation in time of a concentration of mass–energy produces gravitational waves. Therefore, the complete but obviously too long title should have been: 'What are the sources of gravitational waves that can be detected?'. The answer to this question is rather simple: all sources that are sufficiently 'powerful', that is, able to release sufficiently large quantities of energy in the form of gravitational waves.

It has been quite some time since I last proposed an equation, and I definitely do not want you to get out of the habit of reading them .... For this reason, I give below an approximate but compact expression for the luminosity, that is, the amount of energy per unit of time, of a gravitational-wave source. This famous expression, also known as the *quadrupole formula*, tells us that the luminosity of a source of gravitational waves is simply given by the following basic properties of the source (note I intentionally ignore any numerical factor):

$$L_{GW} \simeq Gc\left(\frac{M}{R}\right)^2\left(\frac{v}{v}\right)^6 \simeq \frac{c^5}{G}\left(\frac{r_s}{R}\right)^2\left(\frac{v}{c}\right)^6 \tag{8.1}$$

In the formula, M is the mass of the source, R is its characteristic size (for example, the radius if we think of a star or a black hole), and v is the average speed at which it moves. As usual, I have greyed out the parts related to constants, which do not play an important role in understanding the equation.

Looking a bit more carefully at Equation (8.1), we can see that it contains three terms: the first one on the left is simply the definition of luminosity; the second one in the middle expresses the luminosity in terms of mass and speed; the last one on the right instead uses a quantity that we saw in Chapter 6, namely the Schwarzschild radius: $r_s = 2GM/c^2$ (recall Equation (6.1)). First, let's focus on the piece in the middle of Equation (8.1). It tells us that the luminosity is proportional to a quantity we have encountered several times during our journey, that is, the compactness of the source M/R and, in particular, its square. Furthermore, the luminosity is also proportional to the speed of the source (expressed in terms of the speed of light) and, in particular, to its sixth power.

In its simplicity, Equation (8.1) sheds light on many details that had perhaps remained obscure until now. Since most astrophysical structures have very small compactness and move at much slower speeds than that of light, their luminosity in terms of gravitational waves emitted is inevitably minute. Let's take the concrete example of the Sun. We have seen its compactness is of the order of $M/R \simeq 0.000002 \sim 10^{-6}$ (recall Equation (4.3)) and that it darts through our galaxy at a speed of around 800,000 kilometres per hour, a few thousandths of the speed of light. Therefore, we can use Equation (8.1) to calculate its gravitational-wave luminosity and find that it is: $L_{GW} \simeq 10^8$ watts $\simeq 10^{-8}$ L_{EM}. In other words, the Sun's 'luminosity' in gravitational waves is less than one billionth of a billionth of that emitted in the form of electromagnetic radiation (EM). And if this is true for our star, which still moves at a considerable speed and has a certain compactness, you can easily guess how futile it is to try to generate detectable gravitational waves here on Earth. For example, imagine that you want to build a gravitational-wave source by using a mass of one tonne (1,000 kilograms) and a size of one metre that moves at 100 kilometres per hour. This could be, for instance, a small but heavy car travelling on a closed circular circuit with a radius of one kilometre. In this case, the amplitude of the gravitational waves produced would be as small as $h \simeq 10^{-43}$: a few millionths of a billionth of a billionth of what the best detectors can measure today. I don't think it is worth trying this out....

Instead, let's now focus on the third term in Equation (8.1). It tells us that the gravitational-wave luminosity is greatest when an object has dimensions comparable to its Schwarzschild radius and moves at a speed very close to that of light. This makes perfect sense and explains why the first measured signal of gravitational waves was that produced by a binary system of black holes; they are very simply the most powerful sources imaginable. By definition, the components of the system have a size that is *exactly* equal to the Schwarzschild radius, and they move at about two-thirds of the speed of light in the final stages of their inspiralling orbits. Furthermore, since they are part of a binary system, they offer the greatest possible deviation from spherical symmetry.

In summary, Equation (8.1) gives us a very clear message: if we want to have any hope of detecting gravitational waves with present interferometric detectors, we must collect the signal from objects that are extremely compact (i.e., with $M \simeq R$), have dimensions comparable with the Schwarzschild radius (i.e., with $R \simeq r_S$) and that move at a speed close to that of light (i.e., with $v \simeq c$). Gravitational waves emitted by sources of this type, and placed at a distance of a few hundred megaparsecs, will reach Earth with an amplitude $h \simeq 10^{-21}$ that, although at the limit of our instruments' sensitivity, can in principle be measured.

The Zoology of Gravitational-Wave Sources

Now that we know what it takes to produce strong emissions of gravitational waves, it is useful to classify the various gravitational-wave signals we can detect based on the sources that produce them. As we will see, different sources will imprint their specific properties on the characteristics of the signals, which will vary in terms of amplitude, frequency, and duration in time.

1. *Impulsive signals*. These are very short signals, that is, of the order of tens of milliseconds or less, and they can be produced by different sources. For example, signals of this type are expected

in association with supernova explosions or the gravitational collapse leading to the formation of either a neutron star (as with supernovae) or a black hole. Since these are short and evanescent signals, their 'spectral properties' (the window in frequency in which the signal appears and the amplitude at the various frequencies) are not well defined. In general, however, we are looking at signals at high or very high frequencies, starting from a few hundred hertz and going up to several kilohertz. Given the sensitivity of current detectors and the signal from a supernova being rather weak, impulsive signals of this type are mostly expected from our 'neighbourhood', that is, within our galaxy. Although very common and observable electromagnetically, even at the greatest distances, extragalactic supernova explosions are too far away and will not be 'visible' by gravitational-wave detectors. Studies of the rate of supernova explosions within our galaxy indicate that we should expect one supernova explosion every 30 or 40 years. Hence, it is not surprising that the LIGO and Virgo detectors have not yet measured such signals, although we are all anxiously awaiting the first one!

Precisely because of their 'impulsive' nature, signals of this type could be linked to sources that have been predicted but never observed so far, such as cosmic strings, which are relics of phase transitions in the early universe. Furthermore, should a wholly new and unexpected source exist that no theory has yet predicted, it is very likely that the corresponding signal will be of the impulsive type. For this reason, measuring the first impulsive signal would be a doubly exciting event!

2. *Periodic signals*. These are signals from sources that produce a gravitational-wave signal at a single precise frequency or at a limited set of very specific frequencies. Notably, the amplitude of the signals and the corresponding frequencies are expected to remain constant over periods of tens if not hundreds of years. Potential sources of periodic gravitational waves are clearly pulsars, objects already encountered in Chapter 5, which we know emit a beam of electromagnetic radiation that hits us like the light from a lighthouse beacon. These rotating neutron stars

can deviate from perfect axial symmetry due to small deformations on their surface (our 'mountains' in Chapter 5) , produced when the neutron star's crust solidifies in the rapid cooling process after its birth. Alternatively, the deformations can be present in the stellar core if it is a crystal, or they can be induced by the presence of a strong magnetic field, which, through magnetic tension, gives rise to non-axisymmetric stresses.

In essence, given the great compactness and rapid rotation of neutron stars, it is sufficient that a slight asymmetry is present for it to generate a gravitational-wave signal that is detectable within our galaxy. You may think that this type of signal should be easy to detect. Unfortunately, it is quite the opposite for a number of reasons. Although the emission occurs at a precise frequency, the frequency is unknown. Also, the signal is extremely weak, and the signal-to-noise ratio can even drop below unity. It is, therefore, only through observations lasting weeks or months that it is potentially possible to 'extract' this weak signal from the sea of detector noise in which it is immersed. The observation campaigns carried out so far by LIGO and Virgo have not yet revealed any source of periodic gravitational waves. However, they have provided tight constraints on the 'height' of the mountains possibly present on the surface of some targeted neutron stars, which cannot exceed one millimetre and, in many cases, are well below a tenth of a millimetre.

3. *Stochastic signals*. This type of signal is markedly different from those described so far in that the sources responsible for it are not 'resolved' individually. In other words, it is possible to associate the measured signal only to a collection (often very large) of sources that cannot be distinguished individually by the instruments. Furthermore, each of these sources is too faint to be detected in isolation but, when taken together, they become measurable. To better understand how this is possible, let me use a very simple but effective example. Take the image of a distant galaxy that you can already see with a small telescope. It will appear as a small nebula because the light that reaches you is the combination of the light of tens, if not hundreds, of

billions of stars, which, of course, you cannot distinguish individually. Taken in isolation, none of the stars contributing to the image would actually be visible. However, once put together, their emission combines positively and allows you to see the distant galaxy. The same logic applies to sources of gravitational waves that are not resolved individually, but that can be detected by combining the weak signals that all the sources produce.

There are several classes of sources of stochastic signals. One of these is the stochastic 'background' of astrophysical origin produced by a large number of binary systems of white dwarfs or brown dwarfs. These types of objects represent stars in the final stages of their life, and there are billions of them in our galaxy alone. Brown and white dwarfs are more compact than ordinary stars but also much less compact than neutron stars. Apart from those binaries that are very close and well known, the signal they produce is too weak to be identified individually. The simple reason is that they are not compact enough and do not move in their orbits at sufficiently high speeds (recall Equation (8.1)). However, when millions of these sources are 'switched on' and emitting gravitational waves, the signals are, in fact, detectable as an indistinct background, as long as the observation time is long enough. The typical period of the gravitational waves produced by these sources is about one hour, a frequency too low to be measured by interferometers such as LIGO or Virgo. However, they will certainly be 'visible' by the Laser Interferometer Space Antenna (LISA), which is due to be launched around 2035, and for which they will represent a source of noise.

Another type of stochastic background is of cosmological origin and could have been generated soon after the Big Bang due to very small but non-zero perturbations in the density distribution across the universe. These cosmological gravitational waves are at extremely low frequencies, but a part of them should have a signal with periods of tens of days and, therefore, potentially be accessible to LISA. Since they were emitted only 10^{-43} seconds after the Big Bang, measuring them directly or indirectly (through the 'footprint' they may have left on the cosmic microwave background[4]) would be extremely valuable. It would allow us to obtain information on the physical conditions of our universe in the very first moments of its life.

4. *Signals from binary systems*. These are the gravitational signals 'par excellence', as they are among the strongest possible and the only ones detected so far, either directly or indirectly. Several aspects of this type of signal merit detailed discussion, and I will do this in the following pages. Here though, I will limit myself to remarking on the bare minimum.

In particular, I will call to mind that there are two classes of astrophysical objects that are undisputed champions of compactness, neutron stars and black holes. Hence, they are potentially strong sources of gravitational waves just because they are so compact. To this significant advantage, you can add that it is relatively common to find these compact objects in binary systems composed of two black holes, two neutron stars or in a 'mixed' configuration. This characteristic is doubly important as it gives the source a considerable deviation from a spherical symmetry, in fact, the greatest possible. At the same time, it allows the components of the binary system to reach speeds close to that of light when they are about to merge. Combining these factors, and bearing in mind Equation (8.1), it is not difficult to conclude that those generated by binary systems of black holes and neutron stars are the strongest gravitational-wave signals.

As an effective way of concluding this section on the zoology of gravitational-wave sources, Figure 8.4 provides an overview of what we have discussed so far. The position of each source on the horizontal axis serves to give an idea of the 'characteristic frequency' (or period) at which the gravitational wave is emitted. This, therefore, allows us to compare sources with periods of days (i.e., with frequencies of tens of millionths of a hertz, as for binary systems of supermassive black holes) or of tens of milliseconds, and even less (i.e., with frequencies of hundreds or thousands of hertz, as for binary systems of neutron stars). The vertical position of each source, on the other hand, gives a rough idea of the 'characteristic amplitude' of the gravitational waves.

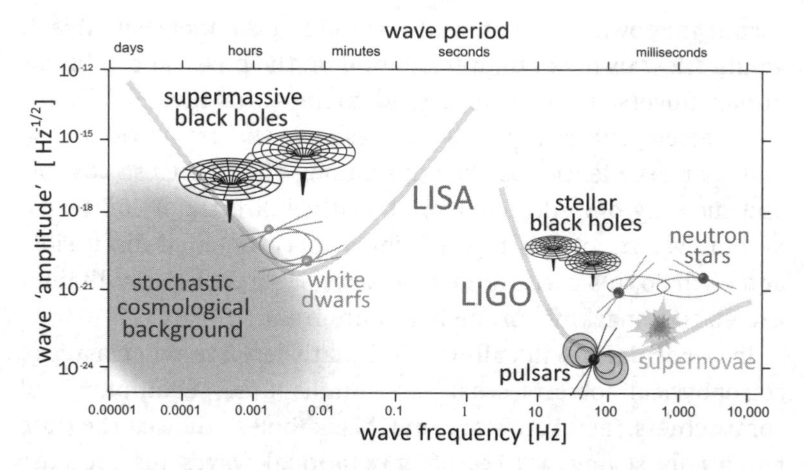

Fig. 8.4 The various sources of gravitational waves identifiable by detectors such as LIGO and Virgo or LISA.

Finally, the lines in the graph represent the sensitivity curves of interferometric detectors such as LIGO and Virgo or the future LISA spatial interferometer and inform us how well these sources can be detected.

Note that the sources shown are not all at the same distance from us. On the contrary, the reported characteristic amplitude refers to sources at the limit of the *observational horizon*, that is to say, the volume of the universe within which each of the sources can be observed. This horizon will be limited to our galaxy (i.e., to a distance of about 100 kiloparsecs from us) in the case of the gravitational-wave signal from a supernova explosion. However, it will be of tens of megaparsecs in the case of binary systems of neutron stars and even hundreds of megaparsecs in the case of binary systems of stellar black holes. Finally, the observational horizon of LISA will include a large part of the observable universe since it will essentially 'see', albeit with different intensities, all of the binary systems of supermassive black holes that enter its frequency range.

Figure 8.4, therefore, brings together every source that we realistically expect to be able to observe in terms of gravitational waves. In reality, deep down, each of us working in this field hopes that at some point a signal will be measured that does not correspond to any of those reported in Figure 8.4. Detecting such an unexpected source would indicate something is wrong in our theoretical description of gravitational-wave sources. At the same time, however, it would give us an excellent reason to question everything!

The Difficult Life of a Couple

We tend to associate the first detection of gravitational waves from a binary system with GW150914 (the first event measured in 2015 by LIGO), corresponding to the merger of a binary system of black holes. However, in reality, thanks to work that began 20 years earlier, it was already evident in the 1990s that Einstein was right, and gravitational waves are produced by a binary system of compact objects.

But let's proceed in order. We have seen that the first pulsar was discovered in 1967 and that, since then, neutron stars have been observed either as isolated pulsars or in binary systems, where the neutron star emits X-ray radiation from the accretion of matter coming from the companion star. Since more than 50 percent (and possibly as much as 85 percent) of stars are not isolated but gravitationally bound in binary (or triple) systems, it did not take long to conclude that two neutron stars could be present in a binary system.

Probably inspired by considerations of this kind, in the early 1970s, the American astronomers Russell Hulse and Joseph Taylor started a systematic campaign of observations of pulsars potentially belonging to a binary system. They used the Arecibo radio-telescope in Puerto Rico, made up of 38,000 aluminium panels housed in a natural cavity, creating a radio-reflecting dish nearly 300 metres in diameter. They soon concentrated their observations on the pulsar PSR B1913+16, discovered in 1974.

On paper, this pulsar is nothing peculiar, with a mass of 1.44 solar masses and a rotation period of about 60 milliseconds, which means that it rotates on itself 17 times per second. There was, however, something anomalous in its radio emission, something not seen with any of the pulsars known at that time. In order to explain this singular behaviour, it had to be acknowledged that the star orbited a companion; the problem was that such a companion should have been visible, but it was not. However, if the companion was also a neutron star, this would explain the observations quite well. This was how Hulse and Taylor came to discover the first binary system of neutron stars, also known as the *Hulse–Taylor binary pulsar*, since it is a binary containing two neutron stars, one of which is a pulsar.

Today, thanks to the enormous precision of radio-astronomical observations of pulsars, we know PSR B1913+16 exceptionally well. For example, we know that the masses of the two neutron stars are very similar (the primary has 1.441 solar masses, the secondary 1.387) and that the orbital period is 7.75 hours. The orbits are elliptic, so the two stars orbit the common centre of mass, and their separation varies continuously, reaching about 1.1 solar radii (746,600 km) at the point of minimum separation (periastron) and about 4.8 solar radii (3,153,600 km) at the point of maximum separation (apoastron). Given the masses involved and the high orbital speeds reached by the stars, the binary pulsar PSR B1913+16 represents a system in which relativistic effects are amplified and therefore easier to measure. For instance, just like Mercury‘s orbit, that of the pulsar is also not closed but experiences a change of the periastron (recall Figure 2.1). The crucial difference is that the periastron of PSR B1913+16 advances by 4.2 degrees every year, while Mercury's advances by only 0.4297 arcseconds per year. In other words, the orbit of PSR B1913+16 varies in a single day as much as Mercury's does in one century!

The most important relativistic effect in PSR B1913+16 is linked to the emission of gravitational waves, which reduces the energy and angular momentum of the binary and forces the two stars to get closer. In particular, their separation is reduced by three millimetres after each period, so, over one year, it has decreased by about 3.5 metres. In absolute terms, the loss of energy from a binary system such as that of PSR B1913+16 is enormous; furthermore, this system is actually in our galaxy

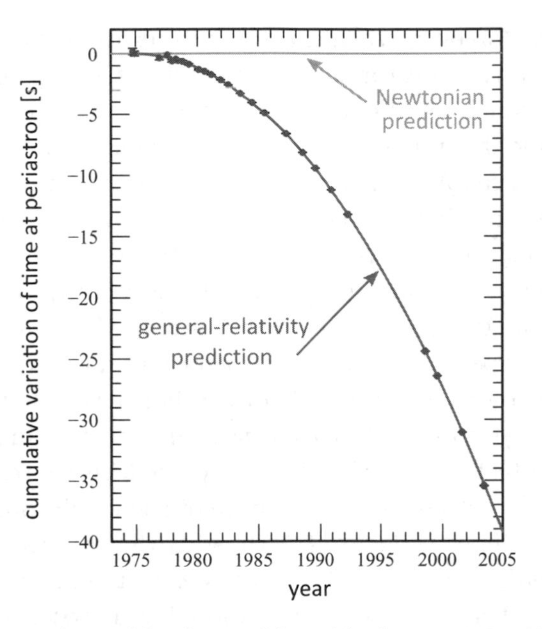

Fig. 8.5 Observations of the decay of the orbit of PSR B1913+16 in terms of the advance of the periastron time. Data by J. M. Weisberg and J. H. Taylor, *ASP Conference Series*, 328, 25 (2005).

and not very far from us at 21,000 light years away. Yet, the amplitude of the gravitational waves produced by the binary is too small for detectors, such as LIGO and Virgo, designed for systems at much higher frequencies (see Figure 8.4). Nevertheless, the emission of gravitational waves is 'visible' through the effects it produces on the evolution of the pulsar's orbit and, in particular, on the change of the time at which PSR B1913+16 should reach its periastron.

Figure 8.5 shows the advance in time of the binary's arrival at the periastron as a function of the year the observations were carried out. That is to say, the decrease in the time required for the pulsar to complete a full orbit is reported with points and error bars in Figure 8.5. Note that the reported data summarise observations carried out over 30 years and clearly illustrate how the orbital period of PSR B1913+16 (approximately 7 hours and 45 minutes) has decreased by almost 40 seconds over such a timescale. Moreover, when expressed in terms of the separation between the two neutron stars, this has decreased by 105 metres!

This example highlights an important feature of binary systems and the title of this section. Independent of their nature and mass, a couple of astrophysical objects that are gravitationally bound in a binary system must, eventually... 'collide'! In the vast majority of cases, the timescales for this process are simply enormous, much longer than the age of the universe. For example, in the case of the Earth–Moon binary system, the merger is expected in 65 billion years and does not represent an urgent source of concern. On the other hand, the timescales for binary systems of compact objects are much reduced and of the order of a few tens of millions of years. This shorter window allows us to 'witness' the occurrence of such collisions.

The solid lines in Figure 8.5 show us the theoretical predictions according to Newton's theory of gravity (just a horizontal line since there can be no variation for a closed orbit) and according to Einstein's general relativity. In the latter case, the curve is not horizontal because the orbit must change (shrink) due to the emission of gravitational waves. It is remarkable how perfectly the experimental measurements fit with the theoretical curve predicted by general relativity. For this reason, the observation of PSR B1913+16 represents the first *indirect* proof of the existence of gravitational waves. Following the measurement of this process, Hulse and Taylor were awarded the Nobel Prize in Physics in 1993. Obviously, this result does not detract from the work of LIGO, whose observation of GW150914 has provided the first *direct* proof of the existence of gravitational waves.

If what I've just told you about the remarkable properties of PSR B1913+16 blows your mind and pushes you to imagine two stars with a diameter of about 25 kilometres orbiting one another at millions of kilometres per hour, then you had better be prepared – there is, in fact, a binary pulsar system that is even more extreme! In 2003, using the radio-telescope at the Parkes Observatory in Australia, an international team, coordinated by the Italian astronomer Marta Burgay, discovered a binary system of double pulsars, that is, a binary in which both neutron stars are pulsars. The system is known as PSR J0737–3039, and the sheer fact that the binary contains two pulsars allows for the extraction of even greater amounts of information. The separation at the apoastron in PSR J0737–3039 is comparable with the solar radius (about 800,000 km), and the orbital period is as short as 2.45 hours! This last property is particularly important. We have seen that the orbits of the two stars are slightly different from those predicted within Newton's theory of gravity. These differences are contained in a number of parameters, known as *post-Keplerian parameters*, which serve precisely to measure how much the orbit predicted by Einstein differs from that predicted by Newton. The measurement of these parameters allows for direct testing of gravity and verification of the correctness of Einstein's predictions. Because PSR J0737–3039 has such a short orbital period, it gives us the opportunity to collect data much more quickly. In fact, the same tests carried out with PSR B1913+16 over a 30-year period took less than a third of the time to perform with PSR J0737–3039. Therefore, it is no coincidence that, only a few years after their discovery, we know many of the details of this fantastic system with incredible precision. More precisely, we know: the mass of the two stars (1.3381 and 1.2489 solar masses), their age (210 million years for the most massive star, 50 million for the less massive) and their rotation periods (a short period of 22.7 milliseconds for the oldest and a 'very long' period of 2.77 seconds for the youngest, for this reason, also called the *lazy pulsar* . . .).

I will conclude this section with an answer for those who may have this question buzzing around in their heads. Given that the separation between the two stars decreases continuously as a result of the emission of gravitational waves, when will the two neutron stars in these systems merge? Well, this may be a disappointment, but you do not need to mark it in your calendar just yet. Indeed, the merger of PSR B1913+16 is expected in about 300 million years, while that of PSR J0737–3039 is expected a little bit earlier, that is, in about 85 million years . . . give or take a day or two . . . .

1 + 1 = 1: The Bizarre Arithmetic of Binary Black Holes

We have already established that binary systems of black holes, neutron stars and mixed systems are the most intense sources of gravitational waves. At least in broad terms, the genesis of these systems is quite clear.

First, as confirmed by abundant observational data, stars in binary systems are more common than those in isolation. Furthermore, it is clear there are binary systems composed of very massive stars that will be able to produce, through a supernova explosion at the end of their life, neutron stars or black holes depending on the mass of the core at the time of collapse. Therefore, it is not difficult to imagine an evolutionary track for the binary system in which the most massive star (the primary) evolves faster and gives rise to a supernova explosion first, thus becoming either a black hole or a neutron star. Assume the conditions of the explosion are favourable. In that case, particularly if the recoil generated by the explosion is not so large as to destroy the binary system, we will obtain a binary composed of a compact object (black hole or neutron star) and a massive ordinary star (the secondary). The latter will move along its evolutionary path, which will be slower, as it is less massive than the primary, until a second supernova explosion takes place, whose remnant will once again either be a black hole or a neutron star. This second explosion also represents a danger for the integrity of the binary system, but it is clear that at least a small portion of these binaries will survive 'intact' and give rise to binary systems of compact objects, the ideal sources of gravitational waves.

What happens at this point? To answer this question, we will first concentrate on the evolution of a binary system composed of two black holes, which we can then split into three distinct phases: quasi-circular orbiting; rapid inspiralling; and plunge, merger and ringdown.

In the first phase, that of quasi-circular orbiting, the two objects move along trajectories that would be exactly circular were it not for the very small variations linked to the loss of energy and angular momentum through gravitational waves. Note that even if the orbits were initially elliptic, as for the binary system PSR B1913+16, they would have become quasi-circular at this point of the evolution. This is because the intense emission of gravitational waves tends to reduce their ellipticity and 'circularise' the orbits. The gravitational radiation emitted in this phase can last millions or even tens of millions of years, depending on the initial separation and the mass of the components. It is essentially periodic and extremely weak since the speeds involved are not high. Furthermore, these waves are emitted at frequencies of the order of a millionth of a hertz, making them impossible to measure by current interferometers.

In the second phase, that of rapid inspiralling, the components of the binary are destined to approach one another inexorably in an increasingly frenetic dance. The orbits become gradually less circular and increasingly have the appearance of a spiral. At the same time, as the separation between the two objects decreases, gravitational waves grow in both amplitude and frequency. Reminiscent of Ravel's *Bolero*, the approach proceeds in a crescendo of rhythm and intensity. This increase in the frequency and amplitude of the gravitational waves, soon before the merger, is also known as *chirp signal* since it resembles the way birds modulate their song.

In the last phase, which includes some orbits before the black holes merge or *coalesce*, the gravitational waves reach their maximum frequency and amplitude, only to be quickly damped out as soon as the coalescence occurs. In the case of stellar black holes, the frequency of the signal in this final phase is of the order of hundreds of hertz (i.e., tens of thousandths of a second in terms of the period). However, it is much lower, of the order of one-hundredth of a hertz (with a period of about a quarter of an hour), for supermassive black holes with a mass of around one million solar masses. In addition, during the last orbits before coalescence (the exact number depends on the specific configuration of the binary system), something quite impressive happens: the emission of energy is so copious that the amount of energy lost in these last orbits is comparable to that lost since the formation of

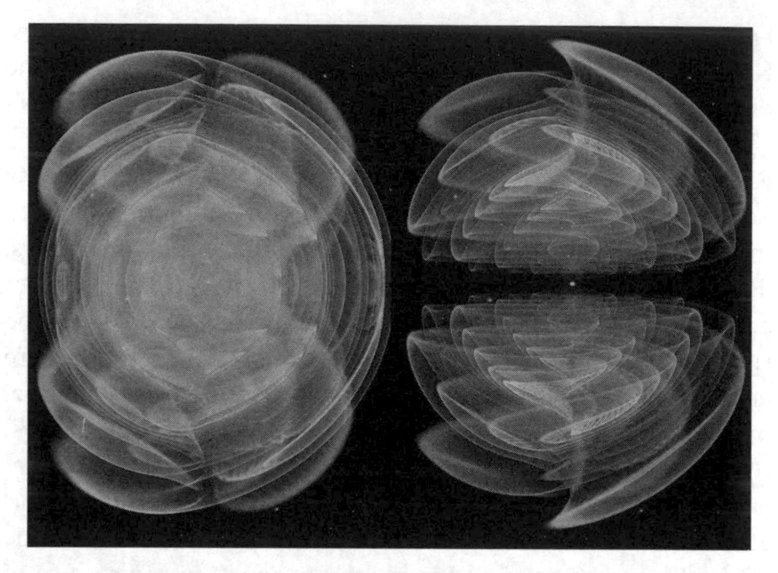

Fig. 8.6 Gravitational waves produced by the merger of a binary system of black holes. © M. Koppitz/C. Reisswig/LR. A black and white version of this figure will appear in some formats. For the colour version, please refer to the plate section.

the system. In the case of stellar black holes, this phase lasts much less than a fraction of a second and, yet, the energy lost is precisely comparable to that given off by the binary in millions of years. An example of the complex geometric structure of the gravitational waves produced in this final stage of the binary is illustrated in Figure 8.6, which shows the signal emitted by a coalescing system of two black holes. In particular, the figure illustrates the gravitational waves emitted in the two polarisations: 'plus' (on the left) and 'cross' (on the right), and the final black hole produced at the centre of each group of waves.

At the moment of the coalescence, a single *apparent horizon* surrounds and encompasses the two, still detached, black holes.[5] The formation of a common and all-embracing apparent horizon takes place when the two black holes are so close to each other that their respective horizons are about to touch. Because of the appearance of this common horizon, a hypothetical external observer would never 'see' the two black holes touch each other (obviously,

I refer to the regions *close* to the horizons since the latter are not visible by definition). This explains the enigmatic title of this section, which jokingly summarises the whole evolution process: 1 (black hole) + 1 (black hole) = 1 (new black hole).

The Perfect Sources of Energy

Interestingly, the new black hole produced by the merger will have a mass comparable to the sum of those of the two initial black holes but a little smaller. The difference is lost in gravitational waves since the energy they carry is completely equivalent to a mass, as the notable Equation (3.2) reminds us. More interestingly, this mass can be a good fraction of that of the two black holes. For example, in the case of the GW150914 event, the LIGO detectors measured the gravitational waveform of a binary system of two black holes with 36 and 29 solar masses, which merged, as a result of the coalescence, to produce a new black hole of about 62 solar masses. The missing part (about three solar masses!) is the mass–energy radiated in the form of gravitational waves in the last milliseconds. If you do the maths, you will then find out that the binary system has actually lost 5 percent of its mass in gravitational waves or, if you like, in distorting the rigid fabric of spacetime around it. The rate of mass loss depends on the properties of the binary system (mass ratio and spin of the black holes) and can be as high as 10 percent in the case of maximally rotating black holes. Now, suppose you think of the energy efficiency of nuclear fission processes, such as those used in nuclear power plants, as being of the order of 0.1 percent. Then it will immediately become clear that the merger of two black holes represents the most efficient way to produce energy in the whole universe. This also explains why these sources are visible billions of light years away.

This enormous loss of energy provides us with an intriguing science-fiction perspective. Perhaps there will come a day when our civilisation is so advanced it can produce energy, not through the collision of heavy atoms, as happens now in nuclear power plants, but through the collision of microscopic black holes in some 'black-hole power plant' located at a safe distance from us in interstellar space. Among other things, such a power plant would solve the problem of waste, simply 'directing' the products of the collision on a path very far from us. Something I haven't told you is that the black hole produced by the merger generally acquires a *recoil velocity* (also referred to as 'kick'). Hence, if the black hole produced is a charged recoiling black hole, it can easily be diverted and directed wherever desired using magnetic fields. This is clearly a science-fiction digression, but it has a correct scientific basis. In any case, don't try this at home!

Let's go back to the real, astrophysical black holes. Within general relativity and, except in very particular situations, the final black hole produced by the coalescence of a binary system of black holes is of the Kerr type, therefore rotating. The rate of rotation and the direction of the spin vector depends on the initial properties of the two black holes, particularly on their mass ratio and their spin properties. However, when it is born, the compact object is not a fully fledged Kerr black hole but rather a 'perturbed' Kerr black hole. What happens at this point is very interesting: the perturbed black hole is able to 'shake off' all the imperfections related to its violent formation, so that, within a few milliseconds (at least in the case of a stellar black hole), every perturbation is radiated away, that is, lost through the emission of gravitational waves. In this way, the newly formed black hole rapidly settles into its final (also called asymptotic) configuration. The gravitational-wave signal emitted when this 'shake-off' takes place has a very special name; it is referred to as the black-hole *ringdown*. Because it decays exponentially fast, it is a very short and evanescent signal, not dissimilar from the sound emitted by a bell when it is gently 'perturbed' by the clapper.

It is important to underline that, just as every bell has a specific sound related to its properties (mass, shape, metallic alloy, etc.), a black hole has a ringdown signal closely related to its properties, namely mass and spin. Therefore, the ability to measure the ringdown signal in the coalescence of a binary system of black holes is essential; it uniquely collects information about the black hole's properties that are not otherwise accessible. Furthermore, it serves to confirm that the product of the merger really is a black hole and not something else. Unfortunately, measuring this signal, whose amplitude is intrinsically evanescent, is anything but easy. Nevertheless, it has already been done, for example, by LIGO and Virgo with GW150914, and this measurement served to exclude the object produced by that merger from being a gravastar.

A Formidable Theoretical Challenge

Perhaps, at this point, you are wondering: 'But how do we know all of these things?' Well, we know because we calculated them numerically almost 10 years before the first gravitational-wave signal was detected. Since the 1990s, a large group of theoretical physicists from all over the world has been developing advanced algorithms and constructing sophisticated numerical codes capable of exploiting the powerful resources offered by parallel supercomputers. This effort has led to the birth of a new branch of physics, known as *numerical relativity*, whose purpose is to solve the Einstein equations numerically (often together with those of relativistic hydrodynamics or magnetohydrodynamics) in order to describe, with the highest possible precision, the evolution of a binary system of black holes or neutron stars. However, this goal is anything but trivial as the equations of general relativity, which describe the dynamics of these objects, are among the most complex known, and the presence of strong non-linearities and singularities makes finding the solution particularly challenging. Indeed, for many years the progress in this area of research has been slow, plagued by repeated failures and frustrations, to the point that, in the late 1990s, no one believed that numerical relativity would ever calculate anything useful.

This belief gave rise to a famous 'bet', made between Kip Thorne (winner of the Nobel Prize in Physics in 2017) and a group of a dozen scientists engaged in numerical-relativity research, myself included. The subject of the bet was who would get a breakthrough first; Kip bet that LIGO would measure the coalescence signal of a binary system of black holes before we theorists could calculate it through numerical-relativity simulations. At stake was a bottle of wine. Undoubtedly, Kip's intent with the bet was to spur us on, and it worked! The first theoretical signals of gravitational waves produced by a binary system of black holes were calculated numerically in 2005. And the first experimental signal was measured 10 years later, in 2015, with GW150914. Kip Thorne sportingly acknowledged the defeat and even mentioned it in his Nobel lecture held in Stockholm on 8 December 2017.

Remarkable progress has been made over the past 15 years through analytical calculations for the early stages of the dynamics of the binary and numerical simulations for the final stages, which are more non-linear and difficult to calculate. This progress has led to binary systems of compact objects now being the best-understood sources of gravitational waves.

Let me stress that being able to compute the gravitational-wave signal has a significance that goes well beyond the ability to solve a difficult mathematical problem. A theoretical knowledge of the signal is, in fact, essential for the actual detection of these sources because, although they release enormous amounts of energy in the form of gravitational waves, they are usually at considerable distances from Earth, so the amplitude of their gravitational waves is minimal, as mentioned several times previously.

Measurements of this type are clearly at the limit of our technology. Therefore, the detection of gravitational waves represents a real 'challenge' not only for experimental physics but also for theoretical physics. When everything is considered, including the strength of the signal emitted, the distance of the source, detector noise, etc., the actual signal measured is often comparable to the background noise of the detectors. Therefore, it is virtually impossible to measure; in technical jargon, we say that, under these conditions, the signal-to-noise ratio is of order unity. However, if the signal were known a priori, it is in principle possible to 'extract' it from the background noise using a technique called *matched filtering*. To get an idea of what this is, imagine you are listening to a very noisy radio broadcast in which, despite the crackling noise, it is still possible to distinguish some words. If we knew some basic information, such as the language used, the topic discussed, the number of voices speaking into the microphone, etc., our brain would be able to 'extract' the signal from the noise and reconstruct what is transmitted almost entirely. What happens with gravitational waves and matched filters is not so different; prior knowledge allows detectors to reveal the faintest signal. Thus, over 20 years ago, the formidable theoretical challenge began to calculate, as precisely as possible, the gravitational waveforms produced by the most common sources (in all their possible configurations) and provide this information to the scientists working on the signal detection and data analysis. To date, this challenge has still not been completed. The prior theoretical knowledge of the signal allows the scientists to compare the experimental signal mixed with the detector's noise, filter out the latter and, thus, find the gravitational-wave signal, even in situations of very low signal-to-noise ratios.

It might seem like a simple goal to achieve, but it is not! Despite the huge progress made since 2005, the solution of the Einstein equations without approximations and symmetries remains extremely challenging. To the high complexity of the Einstein equations, one must add that of the equations of hydrodynamics and magnetohydrodynamics, which are necessary to describe the motion of matter and equally challenging to solve under realistic conditions. As a result, the set of coupled partial differential equations that needs to be solved is so large that its solution is possible only numerically and by exploiting the resources of the most powerful parallel supercomputers. Despite all of these difficulties and problems, the solution of which I have dedicated a good part of my career towards, the prediction of gravitational waves from compact sources, has never been as accurate and precise. All of these efforts have sealed a tremendous synergy between theory and experiments in the common goal of detecting gravitational waves and have shown how far we can go when forces combine.

Closing with Pure Amazement

By now, you must have realised that I can't help but use equations every now and then . . .. Hence, I will close this section by returning for a moment to Equation (8.1) and, in particular, to the expression to the right. I am well aware that what I am doing now directly contradicts what I have claimed so far; I need your attention focused on the constants appearing on the right-hand side of Equation (8.1). Yes, the very same constants I have always recommended you ignore because they are not important. I will actually do more: to make this change of approach clearer, just this once, I will mark in black the physical constants c and G, and grey out all the rest. Hence, for now, the relevant part of Equation (8.1) reads:

$$L_{GW} \simeq \frac{c^5}{G}\left(\frac{r_s}{R}\right)^2\left(\frac{v}{c}\right)^6 \tag{8.2}$$

Now consider a source of gravitational waves with dimensions equal to the corresponding Schwarzschild radius and which moves at a speed close to that of light, just as happens when two black holes are about to merge. In this case, we will have that $r_s/R \simeq 1$ and $v/c \simeq 1$ so the greyed-out parts in Equation (8.2) are roughly equal to one, and the equation effectively depends on the constants only, reducing to:

$$L_{GW} \sim \frac{c^5}{G} \sim 3.6 \times 10^{48} \text{ watts} \tag{8.3}$$

This gravitational-wave luminosity is quite simply humongous! Consider for a moment: there are about one hundred billion (i.e., 10^{11}) galaxies in the universe, and each of them contains one hundred billion stars (another factor 10^{11}), each with a brightness of 10^{26} watts. With Equation (8.2), we have just discovered that, at the moment of the merger, a binary system of black holes radiates as much energy as that emitted, over the same time and in electromagnetic waves, from the entire universe. I don't know about you, but it leaves me in a state of pure amazement whenever I think about this . . ..

The Complex Dynamics of Binary Neutron Stars

The dynamics of a binary system of neutron stars has many similarities with those seen in binary black holes, and some fundamental differences. So, let's arrange things in order and return to the classification of the phases of the dynamics, of which there are generally four: quasi-circular orbiting; rapid inspiralling; merger and post-merger; and collapse and accretion.

The quasi-circular orbiting phase is almost identical to that already discussed for black holes. Both systems can be described with the same equations and the components of the binary treated as point-like particles.

The dynamics of the second phase, rapid inspiralling, is also very similar to that seen for a black-hole system, but not quite the same. Let me explain why. First of all, neutron stars have a larger size than a black hole of the same mass. For example, we have seen that a typical neutron star will have a mass of 1.4 solar masses and a radius of about 12 kilometres, while a black hole with the same mass will have a Schwarzschild radius of about 4.2 kilometres. So, keeping the mass the same, a binary system of neutron stars will merge earlier than a binary system of black holes, simply because the stars will come to 'touch' when the separation between the two objects is of the order of 24 kilometres, rather than 8 kilometres. Second, because neutron stars are made of matter, they can be tidally deformed and, therefore, have slightly different trajectories from black holes with the same mass. The latter point is particularly important, so I'll try to explain it in more detail. We have seen that a non-uniform gravitational field (that is, a normal gravitational field) generates tidal forces that alter deformable objects, such as the mass of water on the surface of the Earth, which can be raised by the tidal field produced by the Moon. A binary system of compact objects generates a non-uniform gravitational field since the curvature in the region between the two objects is greater than that in the more distant regions. Consequently, a deformable object placed in this non-uniform gravitational field will be deformed in shape. Because a black hole represents the limit in the strength of the gravitational field, it is not deformable. We express this fact by saying it has a zero tidal deformability coefficient (also called *Love number*).[6] In other words, we can assume that a spherical black hole will remain spherical even if placed very close to another black hole in a binary system. Here is the crucial difference: this is not the case with a neutron star. If spherical in isolation, a neutron star will assume a deformed, non-spherical shape in the presence of another neutron star.

This change in shape is very significant for two different reasons. The first is that it induces changes in the orbits followed by neutron stars when compared to those followed by black holes of the same mass. Consequently, and all things being equal, the gravitational waves produced by neutron stars during the rapid inspiralling phase will be different from the corresponding gravitational waves produced by black holes. Indeed, it is exactly this different behaviour that allows us to discern (once the mass is known) whether the compact object is a black hole or a neutron star.[7] The second reason is even more important. The tidal deformability of neutron stars (that is, how much they can be deformed in a non-uniform gravitational field) depends on their internal composition. Therefore, in principle, it is possible to determine their equation of state simply by analysing the gravitational waves in the final stages of their inspiral, the point when the two stars are at their closest, and the tidal effects are, therefore, most intense. This is exactly what LIGO and Virgo did for the GW170817 event, which provided the first experimental constraints, albeit still quite uncertain, on the tidal deformability of neutron stars. What happens after the merger can be quite dramatic and, as in the best dramas, the end result will depend on many subtle details that we will explore next.

A Fatal Battle against Gravity

The dynamics of the third phase, the merger and post-merger, is decidedly different from what we have seen in the case of black holes. When two neutron stars merge, they generally do not produce a black hole but a particular type of star, namely, a *hypermassive* neutron star or HMNS. The adjective 'hypermassive' derives from the fact that the mass of this object, which is the sum of the masses of the two initial stars minus that lost through the collision (which we will talk about later), is above the maximum mass that we have seen exist for neutron stars (recall Figure 5.6). This hypermassive star is in equilibrium but in a very precarious equilibrium, which it can very easily lose; for this reason, we say it is in a *metastable equilibrium*. In essence, the gravitational field is so intense (don't forget the hypermassive star is almost twice as massive as the original stars) that it is on the verge of collapsing to a black hole. However, it manages to avoid the collapse, at least temporarily, by rotating very rapidly and in a differential manner, slower in the core and faster in the outer layers. In other words, the hypermassive star produced by the collision fights a desperate battle against gravity, establishing a precarious equilibrium in which it distributes its matter and rotates almost at the maximum possible rate.

Unfortunately for the star, it cannot maintain this precarious balance for long because it rotates so rapidly that it is not even axisymmetric. Instead, the star is deformed and similar to an enormous 'rotating peanut'. In this configuration, however, it loses energy and angular momentum through the emission of gravitational waves. This loss will inevitably lead to gravitational collapse to a rotating black hole. It is not easy to estimate how long the hypermassive star can carry on its battle against gravity. This very delicate balance depends, in a non-linear manner, on a series of factors, such as the mass of the binary, the mass difference between the two neutron stars, the equation of state and the intensity of the magnetic field. Only through numerical simulations is it possible to accurately describe this phase, which is too dynamic and non-linear to be approximated with analytical or perturbative treatments. From these simulations, we have learned that, depending on all the physical conditions mentioned above, the hypermassive star can survive only tens of milliseconds (in the case of very high initial masses) or several hundred milliseconds. Generically, however, a survival time of more than a few seconds seems very difficult, at least for typical neutron stars, with initial masses of 1.3–1.4 solar masses.

During its short life, the hypermassive star performs two essential tasks. The first is to launch large quantities of matter into interstellar space in the form of high-energy winds (a phenomenon we will discuss in more detail later). The second and possibly even more important task is to emit copious amounts of gravitational waves, precisely because it is very like a rapidly rotating peanut and, therefore, has large deviations from a spherical shape. This emission of gravitational waves occurs at precise and almost constant frequencies related to the properties of the hypermassive star (basically the mass and equation of state). The emission is imprinted on what is called the *post-merger* gravitational-wave spectrum, the distribution of the energy lost in gravitational waves as a function of their frequency. The features of the post-merger spectrum are not dissimilar to a fingerprint, so much so that, in principle, it would be sufficient to observe the main frequencies characterising the spectrum to obtain information on the 'holy grail' of nuclear physics: the equation of state of matter at nuclear densities.

Unfortunately, the prospects for the actual observation of this post-merger spectrum are not very good, at least with current interferometric detectors. This is because the frequency of the gravitational-wave signal, which is already very high soon before the merger (around 1 kHz), becomes even higher in the post-merger phase, reaching values of about 2–3 kHz. In practice, the signal measured by the detectors at these frequencies and for sources at typical distances expected for binary neutron stars is far too weak and dominated by noise. As a result, interferometers like LIGO and Virgo are effectively 'deaf' at these frequencies. Exactly this happened in the case of the GW170817 event, which remained visible for the LIGO detectors for some time before the merger, but whose signal soon disappeared because it was at frequencies too high to be measured. Fortunately, these bleak prospects will change radically with the construction of 'third-generation' interferometric detectors, such as the Einstein Telescope and the Cosmic Explorer. These marvellous instruments have arms that can be as long as 40 kilometres and are cooled to extremely low temperatures to attenuate the instrumental noise; more importantly, they are designed to be much more sensitive in the 2–3 kHz region. When these detectors become operative in about 15 years, it will finally be possible to measure this weak but enormously important high-frequency signal.

What happens when the battle is lost? Once the hypermassive star has lost too much of its rotational energy and has, thus, exhausted its ability to withstand gravity, it is forced to collapse into a black hole, leading to the fourth phase of the dynamics of binary neutron stars mentioned above. However, not all of its matter is absorbed by the event horizon of the resulting black hole. This is possible because the hypermassive star is relatively large in size and what collapses to a black hole is only its inner core. When this happens, the outer parts of the hypermassive star are at a considerable distance from the centre, up to 15–20 kilometres. Since the event horizon produced has a radius of about eight kilometres, some of the star's matter ends up on stable orbits and, therefore, is not captured by the black hole. Well, at least not immediately. At this point, the system resulting from the collapse of the hypermassive star is composed of a black hole surrounded by high-density, high-temperature matter orbiting it. This essentially constitutes a geometrically thick accretion disc, also referred to as an 'accretion torus'. To get a better idea of what goes on after the merger, you should visit my web page and look for the section dedicated to the visualisation of the numerical simulations carried out by my research group. I believe that visualisation is better than a thousand words to explain what happens in the complex merger of a binary system of neutron stars....

As you can imagine, the accretion torus is also not destined to last too long. The matter in the torus is highly magnetised and subject to instabilities (particularly the magnetorotational instability), which trigger its accretion on the black hole. Hence, a few seconds after its creation, all of the torus matter is absorbed by the black hole, which remains the only true witness of the entire coalescence process of a binary system of neutron stars. Once again, therefore, the bizarre arithmetic of these systems can be summarised as follows: 1 (neutron star) + 1 (neutron star) = 1 (black hole).[8]

A Mystery That Started with the Cold War

A very important difference between the merger of binary black holes and binary neutron stars is that an electromagnetic counterpart is generally expected from the latter. This expectation is quite natural if you think that the merger of neutron stars amounts essentially to the collision at speeds close to that of light of two massive and compact concentrations of matter. Almost inevitably, a process of this type, with the high densities and temperatures it produces, is bound to produce a signal relating to electromagnetic radiation. However, understanding this signal requires that I tell you about a mystery that began in the 1970s, in the middle of the Cold War.

In those years, the Eastern and Western blocs were heavily involved in fierce competition for both armaments and technological development. For reasons that today seem all too obvious but back then were not, the two blocs had agreed on a ban on all nuclear bomb tests. In this context of non-proliferation of nuclear weapons mixed with the general distrust between the two superpowers, in the 1970s, the United States launched a spy satellite capable of collecting radiation in the gamma-ray band. This satellite was very different from how we may imagine it: instead of pointing to the sky, it was permanently pointed towards Earth, and, in particular, scanning the vast territory of the (then) Soviet Union. The purpose of this satellite was obvious enough: to reveal the emission of gamma-ray signals produced in the case of a nuclear explosion. In other words, it was a satellite built to measure flashes of gamma rays. The paradoxical aspect of this spy story is that the satellite did not record any terrestrial events because the Soviets stuck to the nuclear-test ban. At the same time, the satellite was not a total waste of technology. On the contrary, it recorded a whole series of gamma-ray flashes of great intensity and clear celestial provenance. Since these results were protected by military secrecy, the scientific community knew very little about them. Indeed, it took many years for these observations to be declassified and many more years for the importance of these mysterious sources of gamma-ray flashes to become evident. In fact, we had to wait until the early 1990s for astronomical research in this area to be fully developed.

Today, after about 30 years of observations of the 'gamma-ray sky' (or gamma-ray astronomy), we know that alongside galactic and extragalactic sources, whose properties are known through repeated measurements, there are also two classes of sources that are as interesting as they are mysterious. First of all, *episodic* sources, in the sense that they are observed only once and for very short periods of time. For this reason, they are called *gamma-ray bursts* or GRBs. Since these flashes are very short, episodic, appear without precursors in other bands and cannot be measured more than once, the uncertainty about the sources generating them is still high. However, we do know that they fall into two broad classes: *long* and *short* bursts. Despite the name, even the long bursts are extremely short, lasting between a couple of seconds and about a minute. On the other hand, the short bursts have a much shorter duration, ranging from a fraction of a second up to about two seconds. Of course, it's not always easy to distinguish short bursts, accounting for about 30 percent of the total population, from long bursts. However, when you look at the statistics of all gamma-ray bursts observed so far (roughly 2,300, with a current average of one gamma-ray burst observed every day), it becomes clear that there are two distinct populations. One population has an average duration of around 0.3 seconds for the *short* gamma-ray bursts, and the other of about 30 seconds for the *long* gamma-ray bursts.

In both cases, we are dealing with explosive phenomena that release enormous amounts of energy (of the order of 10^{50}–10^{52} erg $\cong 10^{43}$–10^{45} joules) in a matter of seconds or less. Furthermore, the observations indicate the clear presence of a collimated jet that propagates at relativistic speeds, producing gamma-ray radiation during its propagation or when it collides with the interstellar medium. To get an idea of how enormous the energy released by these objects is, think of it like this: in a matter of seconds, gamma-ray bursts release an amount of energy equal to that emitted by all the stars in our galaxy (amounting to trillions) in one whole year! For precisely this reason, such flashes are visible from enormous distances and even from the most remote corners of our universe. Considering the energies involved, the fact they are generated so far away is actually a good thing, as being hit by even one of these flashes could represent the end for a large part of life on our planet....

It will not come as a surprise to learn that there is no shortage – indeed, there is an excess – of theoretical models capable of explaining the phenomenology observed with gamma-ray bursts in a more or less satisfactory way. Putting aside the most imaginative and exotic hypotheses, today's scientific community agrees in associating the long gamma-ray bursts with the implosion of particularly massive stars. In this scenario, the collapse of the stellar core generates a black hole that accretes matter from the stellar interior in the form of an accretion disc and launches a jet that is powerful enough to drill through the rest of the star and to emerge, becoming visible.[9]

By the end of the 1980s, a plausible explanation was proposed for the short gamma-ray burst involving the coalescence of a binary system of neutron stars.[10] This hypothesis seemed perfectly reasonable from the beginning. The timescales associated with the merger of a neutron star binary are precisely of the order of $0.1 - 1$ seconds, and the binary system has a huge reservoir of gravitational energy. Even releasing a few percent of this energy would be sufficient to account for the measured luminosities of short gamma-ray bursts. Notwithstanding the plausibility of this model, and hundreds of short gamma-ray bursts being observed over the past 30 years, the confirmation of this hypothesis only came in 2017 and through an event we have previously discussed: GW170817.

Much More Than Sources of Gravitational Waves

I have already mentioned that the gravitational-wave signal from GW170817 stopped abruptly at the moment of the coalescence between the two stars. However, I have not mentioned that something amazing happened soon after that. Just 1.74 seconds after it disappeared from the gravitational-wave channel, GW170817 'reappeared' in the electromagnetic channel as a short gamma-ray burst: GRB 170817A! It was possible to put together the arrival times of the gravitational-wave signal at the two LIGO detectors and consider that the Virgo detector had not seen the signal, although, in theory, it could have done. Scientists from the LIGO–Virgo collaboration promptly located the large, but not huge, region of the sky where the signal came from and which corresponded to Virgo's *blind spot* (i.e., the region it was not sensitive to at that time).[11] Hence, a few seconds after the detection of GW170817, dozens of telescopes and satellites pointed in the direction of origin of the gravitational-wave signal, looking for some transient event that could be associated with it. It was certainly not the first time that a similar exercise observational drill had been carried out, as agreements and exchange of alerts are in place between LIGO and Virgo and a series of observatories scattered over the planet, as well as satellites in orbit. However, the important difference from all the previous exercises, which up to that time had involved binary systems of black holes and were inconclusive,[12] is that this time an electromagnetic counterpart, albeit weak, was found! It was indeed the signal of a short gamma-ray burst (named for obvious reasons, GRB 170817A) which appeared quite ordinary except for its brightness, the lowest ever observed in a gamma-ray burst. Today, we think that there is probably nothing peculiar about GRB 170817A, and the reason it appeared so weak is largely a result of our position with respect to its collimated emission. Interestingly, had we not looked carefully for such emissions in the right place and at the right time, we would probably have 'missed' it.

I recall that the observations of gamma-ray bursts, short or long, show the presence of a collimated jet of gamma-ray radiation, which we ultimately receive on Earth. Similar to the case of the radio emissions from pulsars, we can observe a gamma-ray burst only if we are lucky enough and its emission beam intercepts the Earth in its propagation. If we think of the jet as a very thin cone, a bright gamma-ray burst is measurable if our telescopes fall within the opening angle of this cone, which is typically less than 10 degrees. Obviously, we do not observe *all* of the gamma-ray bursts produced, even if they are relatively close to us. On the contrary, we can observe *only* those that are 'pointed' towards Earth, which amounts to only about one percent of the total. As for GRB 170817A and its enigmatic low luminosity, today, we believe it appeared dim simply because, given the relative orientation, we were only marginally invested by the emission cone of the relativistic jet. For this very reason, the signal would, in all probability, have been missed had we not pointed the best telescopes in that direction when it was brightest and searched for a smoking gun. It is only thanks to the tenacity of the astronomers and their desire to find the first electromagnetic counterpart to a gravitational-wave signal that the window of multi-messenger gravitational-wave astronomy was finally opened. A window whose importance we have already discussed. More importantly, the detection of GRB 170817A has undisputedly shown that the hypothesis put forward nearly 30 years before, namely, that the coalescence of neutron stars produces short gamma-ray bursts, was not only reasonable but also correct.

If it is now clear that merging neutron stars produce light in the form of a gamma-ray burst, you may still be wondering: 'How is a relativistic jet actually produced in this process?'. This is an excellent question! After all, it is by no means obvious how two dense, compact objects colliding at nearly the speed of light can produce an intense jet of radiation. Without getting into specifics, I must admit we don't yet know the answer to this question. However, there are helpful clues as to how this can happen, which come from numerical simulations, the only kind

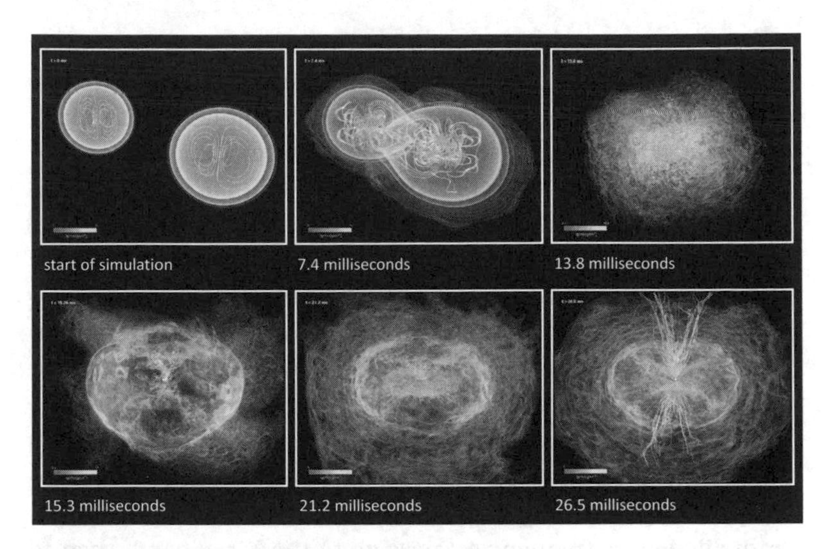

Fig. 8.7 Different phases of a numerical simulation of the coalescence of two magnetised neutron stars. © M. Koppitz/C. Reisswig/LR. A black and white version of this figure will appear in some formats. For the colour version, please refer to the plate section.

that can provide a sufficiently accurate and realistic picture of what happens under these conditions.

An example of these simulations, not the most recent but representing the first study of this kind and performed by my group,[13] is shown in Figure 8.7. The panels illustrate several instants of a numerical simulation of the coalescence of two *magnetised* neutron stars, that is, two neutron stars with an ultra-strong magnetic field.

The simulations, which were state of the art when performed a decade ago, used a number of simplifications and did not come without some limitations. For example, although the geometry of the magnetic field is correct, the velocities of the plasma along the jet are much slower than those measured with short gamma-ray bursts. However, the greatest merit of these simulations was to show, *ab initio* (that is, from first principles), even in 2011, how the presence of a magnetic field is essential to generate a jet from the coalescence of a binary system of neutron stars. In other words, through the joint solution of the Einstein equations and those of relativistic magnetohydrodynamics, these simulations have taught us that it is possible to reproduce the phenomenology observed in short gamma-ray bursts, at least in terms of timescales and energy release, after colliding two magnetised neutron stars. Alas, 10 years later, we are still far from simulating a short gamma-ray burst in a realistic manner. However, the most modern simulations show increasingly clearly that binary systems of magnetised neutron stars represent the most promising way for us to understand how a jet can be produced and move at relativistic speeds towards us, signalling one of the most catastrophic events in the universe.

In the hours and days following the first observations of GRB 170817A, the electromagnetic signal, which initially was obviously in the gamma-ray band, continued to be observed at other, gradually lower, frequencies, passing to the X-ray band, then to the visible band, and finally to the infrared band. Currently, after several hundred days, it is still possible to detect an emission from GRB 170817A in the radio band. This type of signal is called *afterglow* and is produced because, as the source of emission moves away from the coalescence site, it expands and cools. In this way, the emission decreases in intensity and its maximum shifts to ever greater wavelengths. This is not very different from what happens with fireworks when the brightness produced by the explosion fades away and disappears as the cloud of hot gas expands and cools.

An Explosion Worth Its Weight in Gold

In addition to the afterglow, the emission from short gamma-ray bursts can be accompanied by a signal in the ultraviolet and optical bands with a decidedly different origin and which is due to the radioactive decay of material expelled during the coalescence. In fact, a very robust feature of the merger of a binary system of neutron stars is that part of the matter that initially constituted the two stars is ejected to large distances, or, using more technical jargon, 'lost to spatial infinity'. The reason for this is that part of the matter in the collision is highly energetic, with speeds higher than the escape velocity from the gravitational field of the system. The simulations are not yet precise enough to estimate how much mass is ejected; it depends on a whole series of factors related to the properties of the binary system. However, it is quite clear that it oscillates between one-tenth and a few percent of a solar mass. In other words, only a very small fraction of the system's total mass is actually lost to infinity, but this matter has enormous importance, as we will see.

If the idea that a part of the matter of the binary system can be expelled seems obvious enough, what happens to this matter is decidedly less obvious. The first prediction for this was made in 1998 by the astronomers Li-Xin Li and Bohdan Paczyński.[14] Using a simplified model, Li and Paczyński argued that this matter, which is neutron-rich, that is, composed of nuclei with a high number of neutrons (after all, it is the matter that was originally part of the neutron stars), could undergo radioactive decay. As a result of this decay, it would generate an electromagnetic emission (quite generically, radioactive matter emits light when decaying into lighter elements), first in the ultraviolet band and then, over time, in the visible and infrared bands. This electromagnetic signal, whose genesis is very different from that of the afterglow, was named a *kilonova* signal, as it was estimated to be about 1,000 (kilo) times brighter than that emitted by a *nova*.[15] Amazingly, a kilonova signal was observed with GRB 170817A, when a faint decay signal was recorded about 24 hours after the first emission in the gamma-ray band. At first, this signal emerged with a predominant emission in the blue wavelengths (for this reason called a *blue kilonova*) but then faded towards longer wavelengths centred around the red (a *red kilonova*). This was undoubtedly a beautiful firework, but it was also a strong confirmation of Li and Paczyński's prediction. The importance of detecting a kilonova emission is that it provides the final and compelling evidence that a short gamma-ray burst is intrinsically linked to the coalescence of neutron stars. Only neutron stars in a merging system of compact objects can eject matter with the high neutron content necessary to produce and feed the kilonova signal.

But there is more. The faint kilonova light signal tells us not only that a very small portion of the neutron stars can decay radioactively. It also testifies that the merger process distributes into the interstellar space matter containing heavy and neutron-rich nuclei. In this way, it solves another problem that has plagued astronomy for years!

But let's proceed step by step. Throughout their life, we have seen that stars process light elements into increasingly heavier elements through a chain that starts from hydrogen to produce helium, carbon, nitrogen, neon and reaches iron in the most massive stars. Let me also remind you that 'heavy elements' are those with an atomic mass greater than that of iron, which is 56. Examples of heavy elements are strontium (with an atomic mass of 88), caesium (133), europium (152), platinum (195) (197) and, obviously, uranium (238). Stars can generate heavy elements only in very small quantities, given that the main chain of nuclear production is powered by fusion processes that start from hydrogen and end with iron. Furthermore, when these nuclear processes are exhausted and the stars reach the final stages of their evolution, all of these elements (be they 'heavy' or 'light') remain trapped inside the stars, with no possibility of being 'scattered' in the interstellar space. This simple consideration leads to rather complex questions: 'How could a planet like ours, which is composed of very heavy elements, such as platinum and uranium, have been formed? Where did it get its heavy elements from?'.

An answer to these questions was provided in the early 1960s when it was realised that massive stars, whose cores contain elements as heavy as iron, can explode in the form of supernovae and thus distribute these elements in the interstellar space. In addition, both during the actual supernova explosion and in the expansion of the ejected material that follows, *nucleosynthesis* processes can occur, nuclear processes that synthesise heavy elements from the ashes of the exploded star. Hence, these studies show that supernova explosions and the accompanying nucleosynthesis can generate heavy elements such as those mentioned above. From this, it follows that all of the matter present on Earth (including us, since we, in small parts, contain iron and heavy elements) results from the million-year-long work of nucleosynthesis inside the stars. In other words, we are 'sons (and daughters) of the stars', well, at least from a nuclear-physics point of view....

The problem with this idea, which has emerged only over the last 20 years, is that this explanation, however plausible and partially correct, is insufficient to explain the abundance of the heaviest elements measured in the stars around us. In particular, the calculations estimating how many of these heavy elements are produced in supernova explosions indicate a much lower abundance than that measured. In other words, there is too much gold in the universe, more than supernova explosions can produce....

A solution to this problem was first suggested in the mid-1970s, when its severity had not yet become fully apparent. In particular, it was noted that supernova explosions are not the only processes in which it is possible to expel large quantities of neutron-rich material and produce heavy elements. It was pointed out that the coalescence of a binary system containing at least one neutron star is equally effective in this regard.[16] Inspired by this suggestion, a whole series of simulations have been carried out in recent years, which have shown very clearly that the coalescence of a binary system of neutron stars is indeed ideal from this point of view. This is because the temperatures and densities reached in the coalescence are so high, and the matter is already so neutron-rich, that the nucleosynthesis of heavy elements is very efficient. Not only that, the abundance of heavy elements produced in this way is in good agreement with the observations and depends only weakly on the properties of the binary system. In essence, the new paradigm that has been developing over the last decade is that binary systems of neutron stars have the potential to be the optimal 'production site' for heavy elements in the universe.

However, all these considerations were only theoretical predictions and nothing more than the result of complex numerical simulations. For all practical purposes, they were unsupported by any observational evidence. Fortunately for us, the GW170817 event changed all of that! The observation of the kilonova signal has provided confirmation that binary systems of neutron stars eject substantial quantities of matter into the interstellar space and that this material can synthesise heavy elements, as predicted by the theoretical simulations. This confirmation was possible because the light observed in the kilonova signal of GRB 170817A showed traces of the presence of strontium, an essential heavy element. This was one of the smoking guns sought for years and the undisputed evidence that heavy elements are produced in the merger of neutron stars in GW170817!

If you want to grasp how extraordinary this result is, consider that the quantity of heavy elements produced in the nucleosynthesis that accompanied GW170817 is equal to 16,000 times the mass of the Earth. Even more impressive is the idea that a good fraction of these heavy elements is represented by platinum and gold. And approximately 10 'earth masses' (that is, 10 times the mass of the Earth) of platinum and gold have been produced. So, if you are searching for gold, all you have to do is travel in the direction of GW170817, and you will find more than you can carry back.... These considerations also bring with them another lesson. We are, indeed, 'children of the stars', but also 'children of neutron stars'!

I want to conclude this *tour de force* on binary neutron stars with a general consideration. I am fully aware that the amount of information I have given could be overwhelming, if not confusing. However, I can assure you that these systems are as impressive to me as they are to you, and I've been working on them for decades. How could anyone not be amazed to realise that, within only a few hours, GW170817 has provided clear answers to questions that have remained open for decades? With GW170817, we have received confirmation that merging neutron stars are excellent sources of gravitational waves and emit light in the form of short gamma-ray bursts. Furthermore, they also eject a few percent of a solar mass of matter that is particularly rich in neutrons. This matter spreads through the interstellar medium producing most of the heavy elements present in the universe, one of which, gold, we humans seem particularly fond of.

For all of these reasons, the coalescence of a binary system of neutron stars represents a firework of incredible beauty and a fantastic combination of physical phenomena. All the fundamental aspects of modern physics converge in this process: the presence of enormous curvatures of spacetime, the emission of gravitational waves, relativistic hydrodynamics and plasma physics, nuclear physics and the synthesis of elements. This is why I like to call binary neutron stars *Einstein's richest laboratory*.

THE END OF THE JOURNEY

Gravity . . . attracts! This was obvious to you before you started reading this book and is even more obvious now that you have reached the end of it. At the same time, however, I hope you now agree with me that gravity is also *attractive*, which is far less obvious.

This book was written with the intention of showing how attractive gravity is and how it is inextricably linked to what we know about the world around us and the most remote corners of the universe, which are still governed by its laws.

Together, we have accomplished a virtual journey into the world of gravity, as revealed to us by Einstein more than a hundred years ago. I can well imagine it has been an exhausting journey for some of you, forcing you into extraordinary exertions of the imagination and exposing you to concepts that I realise are difficult to accept, at least when encountered for the first time. It was the same for me when, at the age of 20, I started studying this theory by myself. But, don't tell me I didn't warn you that an abundance of imagination and a good deal of patience was needed.

On this trip, we have searched for answers to a series of simple but not trivial questions, which I would like to recall here:

Why does an apple fall from the tree instead of floating in space? What is spacetime? What does its curvature consist of, and how is it produced? Can time be bent? How does a black hole work, and how can we 'build' one? How is it possible to photograph it if it does not emit light? What are gravitational waves, and why are they difficult to measure?

Perhaps these questions still seem absurd and without answers. Or maybe not, maybe now you find them reasonable and even interesting. Finally, perhaps, you have an answer for each of them that makes sense after all and, if that is the case, then the effort made to write this book – and yours, in reading it! – was not in vain.

Our journey together ends here, at least for the moment. When you close this book, I hope it leaves behind a trace of the most important message it has tried to convey: do not hesitate to use your curiosity and ask yourself questions, even questions that may seem bizarre. Let your intuition be your guide, but not your master. As I have shown with countless examples, our idea of how the world 'functions' is inevitably linked to the experience we have of it, which is necessarily limited. What we experience in our lifetime on Earth is but a drop in the ocean of the physically possible.

Finally, and more than anything else, don't limit your imagination. Before advanced mathematics, complex simulations and sophisticated experiments (all of which are indispensable to progress in science) come the agility of our minds and the journeys of the imagination. More than anything else, it is that which allows us to extend the limits of knowledge.

ACKNOWLEDGEMENTS

I considered the idea of writing this book countless times and, just as many times, dismissed it. And always for the same reason: lack of time. The turning point, and therefore the determination to write a popular-science book about gravity, took place aboard a small sailing boat, the *Alea*, in the middle of the Mediterranean while discussing science with my brotherly sailing companion, Luca Bonatti. Faced with such a fine mind and, at the same time, with such an opinionated ignorance, it seemed necessary and inevitable to write a book to expose the rudiments of modern gravity. To Luca, therefore, my first thanks, for having persuaded me to overcome my hesitation.

Thanks must also go to those who made writing this book technically possible: starting with Marina Forlizzi, who believed in it from the beginning, much more than me. To continue, Manuela Galbiati and Rizzoli's team who patiently guided me on a project very different from those I am used to and in a language that, unfortunately, I use only occasionally. I am also indebted to Donato Bini, Carlo Rovelli and Olindo Zanotti for their careful reading in humanly impossible times and to all my students and collaborators, who have patiently waited for the end of this nuisance and to get back to a normal life of research and teaching.

Finally, I must also express my gratitude to Carolin, Anna, Emilia and Dominik, for having endured with stoic patience all of the frustrations that writing a book entails, for the enthusiasm shown towards something that often seemed decidedly useless, and for letting me spend too many weekends in front of a small screen rather than an open horizon.

For this English edition, a special note of thanks must go to Victoria Hislop, who patiently and skilfully copy-edited my English translation.

NOTES

1 Gravity ... Attracts!

1 The radioactivity of our bodies is something unfairly ignored by most. Since we are continuously exposed to gases, liquids and solids that are naturally radioactive, when we consume these substances, we too are radioactive. Radioactivity is not dangerous in itself, inasmuch as there are naturally radioactive elements with which we are often in contact. Of course, radioactivity can become hazardous if its levels are too high. From this point of view, there is no reason to worry, as our radioactivity is so modest it is not dangerous to us or to those around us.

2 The Fathers of Gravity

1 In a reference system in which the two bodies have a position given by the vectors $\vec{r}_1$ and $\vec{r}_2$, their separation is given by the vector difference between the two, $\vec{r} = \vec{r}_1 - \vec{r}_2$, whose magnitude (or modulus) is indicated as $\left|\vec{r}\right| = r$.

2 The second of arc is a unit of measurement of angles used both in navigation and astronomy, and represents an alternative way of expressing the amplitude of the angle span by an arc. In practice, in addition to measuring the amplitude of an angle in degrees, a set of units we are familiar with (but one that is not convenient in navigation or astronomy), smaller units are introduced, namely, the *minutes* of arc and the *seconds* of arc. Therefore, an arc of one degree is divided into 60 minutes (each corresponding to 0.017 degrees) or 3,600 seconds (each corresponding to 0.00028 degrees). Using degrees rather than arcseconds, we would say that Mercury's perihelion changes by only 0.0121 degrees every century!

3 Spacetime, Curvature and Gravity

1 For those interested in a more rigorous mathematical definition, spacetime is defined as a *variety* with events as elements. To this variety, it is possible to associate a 'map' of coordinates through which the invariant distances between events can be measured. This map is described by a rank-two tensor, called a *metric tensor* or simply a *metric*. Therefore, the distance between events is measured by the scalar product, which is mediated by the metric, of the separation between events and is a scalar invariant, that is, the same in all coordinate systems.

2 To be precise, the m in Equation (3.2) indicates the so-called *rest mass*, that is, the portion of mass that does not depend on the object's state of motion. This rest mass has no value in the case of a photon, which thus has a zero rest mass. A more general expression of Equation (3.2), and one that is valid for all types of particles, is $E^2 = m^2c^4 + p^2c^2$, where p is the *linear momentum* of the object. This expression is also valid for a photon, which always has a non-zero momentum proportional to its frequency.

3 For the interested readers, I can briefly add that the Einstein equations summarised in Equation (3.3) are 10 highly non-linear, second-order, partial differential equations. They are written in covariant form through the use of two rank-two tensors, namely the Einstein tensor $G_{\mu\nu}$ and the energy–momentum tensor $T_{\mu\nu}$.

4 Note that the example of the sheet and the bowling ball has a logical flaw: it describes gravity through the effect of the bowling ball, but the latter is *already* subject to a gravitational field. We will explore the question further in this chapter, demonstrating how curvature can be measured even without the need for a 'sheet'. For the moment, it is sufficient to point out that this analogy is perfectly valid. It is important to understand here that the presence of matter or energy produces curvature.

5 If you want to do a little quiz, ask yourself: given two points A and B on the Earth's surface, what path between A and B minimises or maximises the distance? The answer is the great *circle* that unites them: the circle passing through A, B and the centre of the Earth. For example, if we imagine that the two points have the same

longitude, that path would follow the meridian that unites them and that passes through the North and South poles. But, of course, in most cases, going through the poles maximises the distance . . .

4 How to Bend Spacetime

1. As you will remember from Chapter 2, the gravitational constant G is equal to $6.67408 \times 10^{-11} \mathrm{m}^3/(\mathrm{kg\,s}^2)$, where m, kg and s stand respectively for metres, kilograms and seconds, which are units in the International System of Units (SI). It follows that the quantity $GM/(c^2 R)$ does not have units, as they all cancel out. Hence, the quantity $GM/(c^2 R)$ is a *pure number*.
2. The symbol $\oplus$ is used in astronomy to indicate the planet Earth; therefore, for example, $M_\oplus$ indicates the mass of our planet. Similarly, the symbol $\odot$, which we will use shortly, indicates the Sun and any quantity relative to it.
3. The parsec is a unit of length commonly used in astronomy as it allows us to account for the enormous distances encountered outside the solar system. For this reason, kilometres are never used in astronomy, but rather light years and their multiples. I recall that a 'light year' does not measure an interval of time but of space: the space covered by a beam of light when travelling for one year. When expressed in kilometres, this corresponds to 10,000 billion (10^{13}) kilometres. On the other hand, a parsec corresponds to 3.26 light years, that is, about 30,000 billion kilometres. Clearly, it is much easier to talk of one parsec than tens of thousands of billions of kilometres . . .

5 Neutron Stars: Wonders of Physics

1. Two fundamental quantities characterise a wave phenomenon, such as that of electromagnetic waves: the wavelength (i.e., the distance between two 'peaks' in the wave) and the frequency (i.e., the number of peaks arriving in a given interval of time). The wavelength, λ, and the frequency, f, of the electromagnetic radiation are inversely proportional. More precisely, they are related as $\lambda = c/f$, where c is the speed of light.

2 Radiation of this type is also called *soft X-ray*, distinguishing it from radiation with higher energy, which is defined as *hard X-ray*. The hardness of a photon in the X-ray band is therefore measured through its energy (or, equivalently, its frequency). Soft X-rays usually have an energy between 0.1 and 0.3 keV (kiloelectronvolt), while hard X-rays have an energy between 10 and 100 keV.

3 I should recall that first classifications grouped the types of stars into 'colour classes', that is, according to the dominant colour in which their emission spectrum appeared. Hence, stars were ordered from the blue-white of the 'hottest' stars to the orange-red of the 'coldest'. In a decreasing order of temperature, or, equivalently from blue to red, the classes were: O, B, A, F, G, K and M. To help to remember the sequence there is a simple phrase introduced by the American astronomer who worked on this classification, Annie Jump Cannon (1863–1941): 'Oh, be a fine girl/guy, kiss me!'.

4 To fully understand why it was a bizarre phenomenon, it would be good to deepen the question a little. It is natural to expect that the timescale of variability of a celestial source is proportional to its size. In particular, the brightness of larger objects can only vary slowly, while that produced by smaller objects (albeit on an astronomical scale!) can change more rapidly. This relation between size and timescale is the consequence of a fundamental requirement: the source is causally connected. In other words, its changes cannot be arbitrarily rapid but can vary on a timescale (ΔT) which must be greater than the time a photon moving at the speed of light (c) would take to travel through a source of a given size. For this reason, this timescale is called the 'light-crossing' time. At a mathematical level, this constraint translates into a precise formula: if R is the characteristic size of the source, then to preserve causality we must impose that $R < c\,\Delta T$. Since ΔT is measured by the observations, we easily obtain an upper bound for the size of the source R. To give you some reference, the light-crossing time of the Sun is about 5 seconds, while its brightness is subject to variations that occur on a much longer timescale of about 11 years. Hence, the Sun must be smaller than 11 light years, which is obviously correct!

5 I. Shklovsky, *Soviet Astronomy*, 11, 749 (1968).

6 W. Baade, F. Zwicky, *Proceedings of the National Academy of Sciences of the United States of America*, 20, 254 (1934).

7 A. Hewish, S. J. Bell, J. D. H. Pilkington, P. F. Scott, R. A. Collins, *Nature*, 217, 709 (1968).

8 The *atomic mass* is the number that indicates how many neutrons and protons are contained within a certain atomic nucleus. A similar but distinct quantity is the *atomic number*, which instead only counts the number of protons in an atomic nucleus. For example, hydrogen has atomic mass 1 and atomic number 1 because its nucleus is made up of one proton only; helium has atomic mass 4 and atomic number 2 since its nucleus is made up of two protons and two neutrons.

9 Neutrons are fermionic particles, that is, they follow a Fermi–Dirac statistical distribution and are therefore subject to the Pauli exclusion principle. This principle establishes that a quantum state can be occupied at most by a single particle of this type. This very principle is responsible for the degeneracy pressure. Using an analogy that has unfortunately become all too clear these days, it is as if fermionic particles were forced to 'socially distance' and therefore could not be too close to each other. When they do get too close, they generate a lot of pressure that prevents further compression.

10 In quantum mechanics, the wave function represents the state of a physical system of, say, an elementary particle. It is a function of time and position, so it can be used to measure the probability that a particle is in a certain region of space at a given time.

11 G. Gamow, M. Schoenberg, *Physical Review*, 59, 539 (1941).

12 The model I proposed is called 'blitzar' and involves the birth of a black hole from a neutron star that collapses, losing its magnetic field. Numerical simulations of this process show that it would produce a signal very close to that observed, both in duration and in the amount of energy emitted. Obviously, this does not mean that the model is correct or that it really explains the observations: the jury is still out . . . (see H. Falcke, L. Rezzolla, *Astronomy and Astrophysics*, 562, 134 (2014)).

13 F. Pacini, *Nature*, 219, 145 (1968).

14 I have decided to provide an example here that is different from PSR J0437–4715 to clarify that these are not 'special' pulsars.

15 L. Rezzolla, E. R. Most, L. R. Weih, *The Astrophysical Journal Letters*, 852, L25, (2018).

6 Black Holes: Champions of Curvature

1 At least for some time, the name 'black hole' also had a rival in 'frozen star'. The latter was coined to emphasise that when a star collapses and gives life to a black hole, it seems to 'freeze' in time for an external observer. However, the rivalry between the two names did not last long, and the scientific community soon adopted the form we all know today, which is as effective as it is evocative.

2 Conversely, if the black hole had a mass of the order of a few solar masses, the tidal forces (i.e., the gravitational forces generated by the rapid change of the gravitational field between one point and another) would be enormously different even over the length-scale between your head and your feet. Hence, you would be stretched like a rubber band, ending up like 'spaghetti'. This is an experience that I frankly do not recommend, not even as a thought experiment....

3 J. Michell, *Philosophical Transactions of the Royal Society of London*, 74, 35 (1784).

4 R. P. Kerr, *Physical Review Letters*, 11, 237 (1963).

5 R. Ruffini, J. A. Wheeler, *Physics Today* 24, 30 (1971).

6 S. W. Hawking, *Communications of Mathematical Physics*, 43, 199 (1975); S. W. Hawking, *Nature* 248, 3031 (1974).

7 As mentioned, supermassive black holes have gravitational fields that vary very gradually compared to the human scale, or, equivalently, when compared with 'objects' having dimensions of the order of a couple of metres. That's why I chose one for my example.

8 The two characteristics are not necessarily linked: there are black-hole solutions that have an event horizon but do not have a physical singularity at their centre. For this reason, they are called *regular black holes*.

9 P. O. Mazur, E. Mottola, *Proceedings of the National Academy of Sciences, USA*, 101, 9545 (2004).

10 C. Misner, J. A. Wheeler, *Annals of Physics*, 2, 525 (1957).

7 The First Image of a Black Hole

1 Plasma is one of the four states of matter, together with solid, liquid and gas. Matter in this state is characterised by a mixture of ions (i.e., atoms stripped of some or all of their electrons) and free electrons. The plasma state is widespread in astrophysics due to the very high energies that can be reached in the cosmos, for example on a star's surface.

2 H. Falcke, M. Kramer, L. Rezzolla, ERC Synergy Grant *BlackHoleCam: Imaging the Event Horizon of Black Holes*; Grant No. 610058 (2013–2021).

3 The impact parameter is used in particle physics and gravitational physics, to measure the perpendicular distance between a target and the closest point reached by a particle along its trajectory.

4 Obviously, given the spherical symmetry of a Schwarzschild black hole, the unstable circular photon orbit is an entire spherical surface with a radius equal to the radius of the light ring. The concept of the light ring can also be extended to rotating black holes but, in this case, the picture becomes much more complicated. In addition to the distinction between photon trajectories that are *co-rotating* and *counter-rotating* with respect of the black hole, the spherical surface of the light ring needs to be replaced by a region that is only approximately spherical.

5 R. Penrose, R. M. Floyd, *Nature Physical Science*, 229, 177 (1971).

6 The effect just described takes the name of *longitudinal (relativistic) Doppler effect*, to distinguish it from the *transverse relativistic Doppler effect*, which leads to a variation of the frequency even when the motion is in a direction perpendicular to that of the emission of light. However, the transverse Doppler effect does not play a major role in what we are discussing here.

7 In mathematics, all these properties are expressed through the *degree of non-linearity* of the equations, which indicates how a small

variation in one of the properties of the plasma can lead to enormous variations in its behaviour.

8 A positron is the antimatter counterpart of an electron and is, therefore, a particle with the same rest mass of an electron but with a positive charge.

9 There are a number of scenarios that could potentially provide the energy for the acceleration. One possibility is that it comes directly from the black hole's rotational energy, through what is known as the *Blandford–Znajek process* (see R. D. Blandford, R. L. Znajek, *Monthly Notices of the Royal Astronomical Society*, 179, 433 (1977)). Another possibility is that it is instead supplied by the rotational energy of the disc, through what is known as the *Blandford–Payne process* (see R. D. Blandford, D. G. Payne, *Monthly Notices of the Royal Astronomical Society*, 199, 883 (1982)). Finally, there is a scenario in which the acceleration is provided through dissipative processes within the magnetised plasma.

10 T. Karras, S. Laine, T. Aila, *A Style-Based Generator Architecture for Generative Adversarial Networks*, arXiv: 1812.04948.

11 Y. Mizuno, Z. Younsi, C. M. Fromm, O. Porth, M. De Laurentis, H. Olivares, H. Falcke, M. Kramer, L. Rezzolla, *Nature Astronomy*, 2, 585 (2018); H. Olivares, Z. Younsi, C. M. Fromm, M. De Laurentis, O. Porth, Y. Mizuno, H. Falcke, M. Kramer, L. Rezzolla, *Monthly Notices of the Royal Astronomical Society*, 497, 521 (2020).

8 Gravitational Waves: Curvature in Motion

1 A wave equation for a scalar field ϕ can be written simply as $\Box\phi = 0$, where $\Box$ is the *D'Alambert differential operator* (or simply D'Alambertian) and is a synthetic representation of a series of partial derivatives of the second order. For example, in a Cartesian coordinate system in three spatial dimensions $\Box = \partial_t^2 - v^2\left(\partial_x^2 + \partial_y^2 + \partial_z^2\right)$, where v is the speed of propagation of the wave.

2 As mentioned in the text, the description I give of the motion of the mattress is valid at the first order in the perturbation. If we wanted to also include non-linear effects, that is, contributions to the

dynamics of the mattress of order higher than the first one, the mattress would also move in the direction of propagation of the wave, as the latter is partially longitudinal and so transfers a certain amount of linear momentum. In practice, we would also drift slightly away from the person who jumped into the water.

3 Alongside these two linear polarisations, there are also two *circular ones* (one for each direction of rotation) in which the deformation produced by the gravitational wave rotates around the direction of propagation of the wave. Referring again to Figure 8.2, in this case, the *Vitruvian Man* would be stretched and compressed along a direction that rotates over time as the gravitational wave propagates.

4 In the very first moments of its life, the universe was extremely hot and dense, so that matter and radiation were 'coupled', and their temperatures evolved in the same way. When the universe expanded and cooled enough, matter and radiation 'decoupled' and, in terms of thermal evolution, each 'went its own way'. The moment of decoupling was very special as the photons were finally 'free' to propagate without constantly scattering off all the particles around. These photons, which were only produced 380,000 years after the Big Bang, are still visible in what is called the *cosmic microwave background*, a very uniform bath of radiation. If we had microwave-sensitive eyes, the night sky would not appear dark but dimly lit by this primordial radiation.

5 While it is easy to define an event horizon in the case of a static or stationary spacetime, it is much more difficult to calculate it in the case of a dynamic spacetime, such as the one that characterises two black holes soon before merger, or a star before collapsing onto a black hole. In these cases, it is much more convenient to adopt the concept of *apparent horizon* to define the two-dimensional surface within which the light is trapped at any given time. In the case of a static or stationary spacetime, the two types of horizon coincide.

6 The Love number has nothing to do with feelings: it is named after the British physicist Augustus Love (1863–1940), who first defined it in his theory of elasticity.

7 It goes without saying that if the mass of the object is above about 3 solar masses, it will not be a neutron star but rather a black hole.

However, there are observations where the distinction is not so sharp. For example, in the event GW190814, the least massive object has a mass of about 2.6 solar masses. It is, therefore, not clear whether the observations refer to the least massive black hole ever observed or to the most massive neutron star ever detected (see The LIGO Collaboration and Virgo Collaboration, *Astrophysical Journal Letters*, 896, L44 (2020)).

8 Much of what we have seen for binary systems of black holes or neutron stars also applies to a mixed binary system, that is, consisting of a black hole and a neutron star. However, there are also two important differences. The first is related to the mass ratio between the two objects, which can be 10 to 1 or even greater given the characteristic masses involved (the black hole is expected to have a mass between 15 and 25 solar masses, while the neutron star is between 1.3 and 2.3 solar masses). The second difference is that the coalescence of a mixed system cannot lead to the formation of a hypermassive star. In this case, the neutron star is either tidally destroyed and builds up an accretion torus or is even 'absorbed whole' if the difference in mass is very large.

9 See E. Nakar, *Physics Reports*, 442, 166 (2007) for a review of the literature on this subject.

10 See S. E. Woosley, J. S. Bloom, *Annual Review of Astronomy and Astrophysics*, 44, 507 (2006) for a review of the relevant literature.

11 Gravitational-wave detectors are generically *omnidirectional*, that is, sensitive in all directions, but not with the same sensitivity. In fact, there are directions in which the sensitivity is maximum and others in which it is very small, representing the detector's 'blind spots'. Hence, as with a radio antenna, it is possible to define a sensitivity map of an interferometric gravitational-wave detector. In the case of GW170817, the fact that the Virgo detector did not measure the signal was nevertheless very valuable. The knowledge of Virgo's blind spot at that time made it possible to narrow the search area for the electromagnetic signal and, therefore, to discover GRB 170817A.

12 In general, a strong electromagnetic counterpart is not expected from the coalescence of binary systems of stellar-mass black holes. The reason is that these systems represent the final stage of a binary

system that has long since lost all traces of matter, either through the collapse processes that produced the black holes or through the supernova explosions and associated shock waves that dispersed the matter in the interstellar space.

13 L. Rezzolla, B. Giacomazzo, L. Baiotti, J. Granot, K. Koveliotou, M.-A. Aloy, *The Astrophysical Journal Letters*, 732, L6 (2011).

14 L.-X. Li, B. Paczyński, *The Astrophysical Journal Letters*, 507, L59 (1998).

15 In astronomy, it is referred to as a *nova*, a thermonuclear explosion on the surface of a white dwarf, caused by the accumulation of hydrogen due to an accretion process from a companion star in a binary system. The explosion can be visible for several days and reaches a brightness that varies between 10,000 and 100,000 times that of the Sun. These explosions release minute amounts of energy when compared to gamma-ray bursts.

16 J. M. Lattimer, D. N. Schramm, *The Astrophysical Journal*, 210, 549 (1976).

INDEX